elementals

iv. fire

volume iv

Stephanie Krzywonos, editor
Nickole Brown & Craig Santos Perez, poetry editors

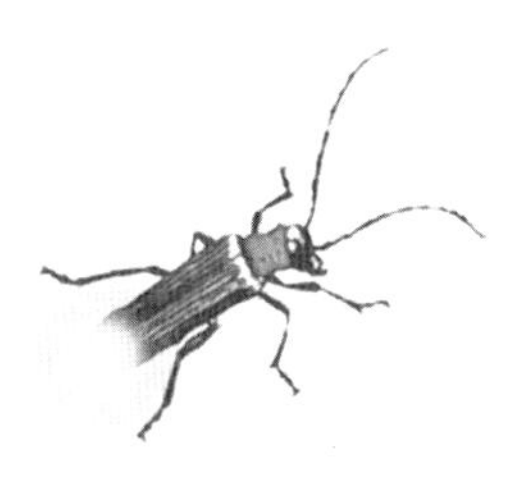

Gavin Van Horn & Bruce Jennings, series editors

Humans and Nature Press, Libertyville 60030

For more information, contact Humans & Nature Press,
17660 West Casey Road, Libertyville, Illinois 60048.
Printed in the United States of America.

Cover and slipcase design: Mere Montgomery of LimeRed, https://limered.io

ISBN-13: 979-8-9862896-3-2 (paper)
ISBN-13: 979-8-9862896-4-9 (paper)
ISBN-13: 979-8-9862896-5-6 (paper)
ISBN-13: 979-8-9862896-6-3 (paper)
ISBN-13: 979-8-9862896-7-0 (paper)
ISBN-13: 979-8-9862896-2-5 (set/paper)

Names: Krzywonos, Stephanie, editor | Brown, Nickole, poetry editor | Perez, Craig Santos, poetry editor | Van Horn, Gavin, series editor | Jennings, Bruce, series editor

Title: Elementals: Fire, vol. 4 / edited by Stephanie Krzywonos

Description: First edition. | Libertyville, IL: Humans and Nature Press, 2024 | Identifiers: LCCN 2024902619 | ISBN 9798986289663 (paper)

Copyright and permission acknowledgments appear on page 148.

Humans and Nature Press
17660 West Casey Road, Libertyville, Illinois 60048

www.humansandnature.org

Printed by Graphic Arts Studio, Inc. on Rolland Opaque paper. This paper contains 30% post-consumer fiber, is manufactured using renewable energy biogas, and is elemental chlorine free. It is Forest Stewardship Council® and Rainforest Alliance certified.

contents

Gathering: Introducing the Elementals *Series*

Gavin Van Horn and Bruce Jennings

Thunderous, cymbal-clashing waves. Dervish winds whipping across mountain saddles. Conflagrations of flame licking at a smoke-filled sky. The majesties of desert sands and wheat fields extending beyond the horizon. What riotous confluence of sound, sight, smell, taste, and touch breaches your imagination when you call to mind the elementals? Yet the elementals may enter your thoughts as subtler, quieter presences. The gentle burbling of clear creek water. The rich loamy soil underfoot on a trail not often followed. A pine-scented breeze wafting through a forest. The inviting warmth of a fire in the hearth.

This last image of the hearth fire is apropos for the five volumes that constitute *Elementals*. The fire, with its gift of collective warmth, is a place to gather and cook together, and not least of all a place that invites storytelling. And in stories the elementals can be imagined as a better way of living still to be attained.

The essays and poems in these volumes offer a wide variety of elemental experiences and encounters, taking kaleidoscopic turns into the many facets of earth, air, water, and fire. But this series ventures beyond good storytelling. Each of the contributions in the pages you now hold in your hands also seeks to respond to a question: What can the vital forces of earth, air, water, and fire teach us about being human in a more-than-human world? Perhaps this sort of question is also part of experiencing a good fire, the kind in which we can stare into the sparks and contemplate our lives, releasing our imaginations to possibilities, yet to be fulfilled but still within reach. The elementals live. Thinking and acting through them—in accommodation with them—is not outmoded in our time. On the contrary, the rebirth of elemental living is one of our most vital needs.

For millennia, conceptual schemes have been devised to identify and understand aspects of reality that are most essential. Of enduring fascination are the four material elements: earth, air, water, and fire. For much longer than humans have existed—indeed, for billions of years—the planet has been shaped by these powerful forces of change and regeneration. Intimately part of the geophysical fluid dynamics of the Earth, all living systems and living beings owe their existence and well-being to these elemental movements of matter and flows of energy. In an era of anthropogenic influence and climate destabilization, however, we are currently bearing witness to the dramatic and destructive potential of these forces as it manifests in soil loss, rising sea levels, devastating floods, and unprecedented fires. The planet absorbs disruptions brought about by the activity of living systems, but only within certain limits and tolerances. Human beings collectively have reached and are beginning to exceed those limits. We might consider these events, increasing in frequency and intensity, as a form of pushback from the elementals, an indication that the scale and scope of human extractive behaviors far exceeds the thresholds within which we can expect to flourish.

The devastating unleashing of elemental forces serves as an invocation to attend more deeply to our shared kinship with other creatures and to what is life-giving and life-nurturing over long-term time horizons. In short, caring about the elementals may also mean caring for them, taking a more care-full approach to them in our everyday lives. And it may mean attending more closely to the indirect effects of technological power employed at the behest of rapacious desires. Unlike more abstract notions of nature or numerical data about species loss, air measurements in parts per million, and other indicators of fraying planetary relations, the elementals can ground our moral relations in something tangible and close at hand—near as our next breath, our next meal, our next drink, our next dark night dawning to day.

For each element, the contributors to this series—drawing from their diverse geographical, cultural, and stylistic perspectives—explore and illuminate practices and cosmovisions that foster reciprocity between people and place, human and nonhuman kin, and the living energies that make all life possible. The essays and poems in this series frequently approach the elements from unexpected angles—for example, asking us to consider the elemental qualities of bog songs, the personhood of rivers, yogic breath, plastic fibers, coal seams, darkness and bird migration, bioluminescence, green burial and mud, the commodification of oxygen, death and thermodynamics, and the healing sociality of a garden, to name only a few of the creatively surprising ways elementals can manifest.

Such diverse topics are united by compelling stories and ethical reflections about how people are working with, adapting to, and cocreating relational depth and ecological diversity by respectfully attending to the forces of earth, air, water, and fire. As was the case in the first anthology published by the Center for Humans and Nature, *Kinship: Belonging in a World of Relations*, the fifth and final volume of *Elementals* looks to how we can live in right relation, how we can *practice* an elemental life. There you'll find the elements converging in provocative ways, and sometimes challenging traditional ideas about what the elements are or can be in our lives. In each of the volumes of *Elementals*, however, our contributors are not simply describing the elementals; they are also always engaging the question, How are we to live?

In a sense, as a collective chorus of voices, *Elementals* is a gathering; we've been called around the fire to tell stories about what it means to be human in a more-than-human world. As we stare into this firelight, recalling and hearing the echoing voices of our living planet, we stretch our natural and moral imaginations. Having done so, we have an opportunity to think and experiment afresh with how to live with the elementals as good relatives. The elementals set the thresholds; they give feedback. Wisdom—if defined as thoughtful, careful practice—entails conforming to what the elements are "saying" and then learning (over a lifetime) how to better listen and respond. Pull up a chair, or sit on the ground near the crackling glow; we'll gaze into the fire together and listen—to the stories that shed light and comfort, to the stories that discombobulate and help us see old things in a new way, to the stories that bring us back to what matters for carrying on together.

Introduction: Tongues of Fire

Stephanie Krzywonos

We only live, only suspire
Consumed by either fire or fire.
—T. S. Eliot, "Little Gidding"

Six years ago in July, I twisted a bright-green leaf of baby spinach by its stem, watching fluorescent light move through it as if it were stained glass, consumed by the miracle of fire. I hadn't seen a plant, other than anemic vegetables, for nine months. I also hadn't seen Sun since the middle of April, when its ember face slid into the horizon—vanishing until the end of August. I was wintering on an Antarctic island and eating lunch in a cafeteria sparsely populated by similarly haggard and pallid people. My hunger for fire was so strong that my eyes felt physically empty. It sounds insane, but I envied that spinach leaf. Days before it had been flown here, the leaf guzzled the sunlight I craved. I even felt a little jealous of the flight crew, who'd witnessed dawn that morning in New Zealand and could again the next day and the next. The inferno whose power I'd underappreciated, the being I once called the sun, had become Sun, and I became its worshiper. In its four-month-long absence, I began to see fire everywhere.

Moonlight gave me a morsel. So did the green auroras billowing across a sky carpeted in stars. The summit of our volcanic island cupped a smoldering caldera, which connected with the burning core of Earth through a magma-filled vein. Each morning, I flicked on the sunlamp next to my computer screen—both forms of fire, and both powered by fossilized Carboniferous plants, fueling the research station, cooking my food, keeping me alive. Sun's fire lived inside the spinach leaf as stored starch, fire that the furnace of my body burned for energy. Rich metaphorical fires smoldered alongside literal ones. The blazing warmth of friendship and the heat of emotions were little flames, like the jealousy of a spinach leaf—a feeling literally powered by my firing neurons, powered by the very same Sun-powered spinach, whose lineage traverses a primordial stellar nursery of gas and dust all the way to the big bang.

In this collection of essays and poems, writers also see fire everywhere and in everything. But what *is* fire?

Western science defines fire as combustion, an exothermic chemical reaction in which material burns rapidly, releasing heat, light, and matter, like smoke and ash. For the writers in this volume, fire is more than combustion. Fire is "a natural force that forms ecosystem processes and forges relationships between the sun and Earth, plants and air, people and places," writes Miriam Morrill, who suggests that fire speaks and that learning its language is one way of conversing with the natural world. "These seeming disparate forms of incandescence are in reality part of the same conversation," proposes Isaac Yuen about the manifestations of Sun, the strange star we think we know. "Perhaps it is better to conceive of these energy permutations as dialects of the solar mother tongue."

Flames and light are only two manifestations of fire. What *else* could fire be?

Fire doesn't only flow. From a time-constrained human perspective, the energy of fire is also stored. "The standing forests, prairies, bogs: they are, yes, carbon banks. But just like me, they also are themselves: sentient, alive, offspring and ancestors, links in a chain of sparks, of sunlight caught and passed along," writes Catharina Coenen, who comes face-to-face with the fire of a coal seam.

Fire is not just our neighbor; fire is part of what makes us human. We are already in relationship with the element: our bodies are fire dependent and fire causing, just like our volcanic home.

Our relationship with fire as a tool is ancient. According to some estimates, it has been a hominid tool for about 1.4 million years, and neolithic humans have had reliable fire-making techniques since at least 7000 BCE. "Fire is a tool the Indigenous people use to ensure the betterment of their livelihood—a tool used for millennia to take care of their landscape, their resources. Fire, like water, is sacred because the plant resources and land calls fire, fire calls wind, and wind brings the water," Ron Goode teaches us. "They all work together, all important to the welfare of one another, important to the sustainability of each one's health."

"Proximity to fire and the wielding of intense heat is also core to many crafts and ancient methods of making: the potter's kiln, the glassblower's torch, the bread maker's oven," writes Nina Mingya Powles on fire living *in* fabric as petrochemical plastic. A tool can become a weapon, and even the modern artisan's crafts can be hard to separate from a warped use of fire.

But humans' use of fire as tool isn't what distinguishes us from other animals. Lightning first showed us how to wield fire. But perhaps our nonhuman kin taught us, too: some plants need to be scorched to germinate, and Australian sparrow hawks use fire to hunt, as Tyson Yunkaporta points out in an exploration of the fire of metabolisms both individual and systematic. Sacred Earth–sky lightning connections parallel belly-brain connections and warped economy-energy connections. "Artificial systems bent to singular goals (like empires, multilevel marketing schemes, celebrity sex tapes, and solar pool heaters) can never harness the economies of scale that are present in nature," he writes. It may be our abuse of fire rather than our use of it that makes us unique among animals.

Fire can calm and be calm, but it is often extreme. The poet Em Strang illuminates fire's paradoxical personality, how it creates as it destroys. On a subatomic level, firelight is both particle and wave, material and not, this "double miracle / of flesh and spirit" offers "unquenchable fire" to our hearts. We are both drawn to fire and repulsed by it. As fiery creatures, we need fire to survive and thrive—yet certain forms of heat and flames are threats.

The day the final essay in this volume drifted into my email, I inhaled my first whiff of wildfire smoke. But the woody scent didn't have the pleasant spicy aroma of campfire smoke. It smelled older, deeper. Not like a single tree felled in the forest, but the whole forest, a Canadian one. I live in Iowa. Last month, July 2023, was the hottest one humans have ever recorded. If fire speaks, what is it saying to us perennially, and especially in an era of climate breakdown and mass extinction?

Life goes on in spite of wildfire—"even in ashfall," as the poet Jane Hirshfield reminds us. Or perhaps life goes on because of fire's churn, and especially in ashfall. "Is this what snow / looks like? How do people / enjoy what seems so perilous?" asks Rina Garcia Chua. Fire is emotionally evocative, fueling both positive and negative feelings. "Safe and convivial fire," writes Glenn Albrecht, inspires reverie and connects us to "comfort, food, entertainment, and security." But "as we edge ever closer to tipping points in the climate, never before experienced emotions are rising from the Promethean depths of our individual and collective psyches. No longer is fire connected to the creation of order; it is now firmly tied to the creation of disorder."

Violent fires coexist with nourishing ones. To David Baker, a blooming tree's flowers are flames, which he describes alongside a suicide bomber: "There is no likeness beyond her body / in flames, for its moment, no matter its moment. / Yet the fringe bloom burns." Fear of violence is "like waiting for the walls to ignite," and in the aftermath of such fires, Camille T. Dungy asks: "Have you ever seen a person walk through the ruins / of a burnt-out home?"

Fire is a cyclical, cosmic force that will not be bent to our control. "Before finding dry wood for the fire— / your gift from the forest— / notice the shape of the hole / and the stones: All life is a circle," affirms Lucille Lang Day. Fire is a gift from Mother Earth, but that gift, "if disrespected, is then withdrawn with dread effects for the world," Jane Caputi warns us. "Firepower is channeled not only into weaponry but also into machines, industry, and technology." Fire is political, economic, and the shapes it "takes in our lives depend on cultural context, including foundational myths and rituals that guide human use."

It is useful to ask not only what fire could be but also *who* elemental fire is—Fire as subject, not object, a thou instead of an it. Instead of trying to control the elements, we should welcome them as teachers and partners in shared regenerative work. In a thought experiment, José González asks: "Is fire alive?" For González, Fire is medicine in ceremony and ritual but has also become a mentor, "showing how sociocultural systems can benefit from periodic and thoughtful regenerative disruption." What else can Fire teach us?

"I vow to be the small flame," says Tamiko Beyer. But how do we "vow to love the fire always"? This is a spiritual, mystical question. Eiren Caffall encounters the mysterious, healing, holy, yet ailing fire of bioluminescence and testifies: "Every small spark is a clue, every shimmer in the black ocean a way to something more luminous and complex."

I hope this volume, each essay and poem a tongue of fire, will be a gathering fire for us. Our relationship with the elements is not a solo endeavor but a communal one. We keep the fire together. "A fire cuts through the blocking vocabulary of the clever rational mind and puts you directly in touch with life," says Charlotte Du Cann, inviting us to gather: "You sit in the darkness with your fellows, human and nonhuman, and feel at home."

Vestment

Jane Hirshfield

For the pear, for the fig,
no difference

effortless   ripening

even in ashfall

The Gathering Fire

Charlotte Du Cann

The people are coming down the street, bearing torches, giant puppets are swaying, firecrackers exploding at our feet. After the procession moves past, we move on to a dark field on the outskirts of town where house-high stacks of wood are set alight, and their flames leap into the night sky. Everyone is whooping and cheering as effigies of politicians are burned by groups in different-colored striped jerseys and a maelstrom of fireworks is unleashed. It is a world turned upside down, total mayhem, and I am laughing. The controls of our lives are bursting open, and we have been suddenly hurtled out of modern times.

This is Lewes Bonfire Night in Sussex: anarchic, vernacular, strange, a remnant of an archaic practice that once rooted us into the earth. It takes place on Guy Fawkes Night in November, when bonfires are lit across the land and people gather round them. Like other "civilized" seasonal events, such as Christmas and Easter, it veers close to the original fire ceremony of Samhain that heralds the winter, the darkening of the year, a hinge time when our ancestors come to speak with us should we be listening.

Something beyond history happens in these topsy-turvy moments—a crack in time, a glimpse into another way of being human. The fire is a bridge.

The Uncivilized Fire

When I look back now, there were always fires: the fire in Spitalfields as we jumped over banging saucepans on Leap Day, just before lockdown. The fire by the Thames that cracked the tarmac open, just after the storytelling company had left. The fire in the Oxfordshire hills, where the road protesters cooked a vast cauldron of soup and sang songs. It was a time of fires and gathering differently, of small lights in jars that were carried across a bridge or lit a path through the woods. Of people pressing together in the darkness, of hearing stories from the ancient world being spoken under stars, of people in masks, of small rituals, of conversations that were out of time, where it felt that the animals and the trees stood behind us. When I waited in a stony alcove altar at winter solstice, with Deepak, my fellow activist, close beside me, the night wind blew through the orchard trees and guttered the candles. Even though we were far away from the fire, you could hear people singing and the fire crackling, and it was then I realized: the ancestors could hear us when we gathered in this way. And in that moment you felt them, and you didn't want to be anywhere else.

And maybe that is why a fire can sometimes be a homecoming in ways you do not expect. Twelve years ago, when I sat in the darkness under the stars and a man in a bear mask came out of the woods and began a folk tale from Siberia, something jolted awake in me. That storytelling fire was at a Dark Mountain Uncivilisation festival held in the Hampshire woods. I had been part of a grassroots community-based movement responding to climate change and fossil-fuel dependency, but in that moment I knew what I had been missing. *Here it was!* People converging in a way that felt ancient but also like the future. They were facing the same existential difficulties, but instead of trying to fix and control them, they were allowing a space for a creative and Earth-based culture to emerge, to make a web of connections that could hold us together when things fell apart.

I changed tracks that night and joined the band of writers and artists that is the Dark Mountain network.

In the following years, we held fires across the land, gathering at strategic points in the year: in Newcastle upon Tyne, in Reading, in the Cumbrian and Cheshire hills, on Rannock Moor and Dartmoor. There were book launches and performances, stories and songs and laughter, and bottles of whisky passed around. When the pandemic came, we asked ourselves: How can we meet when we are pushed apart from one another? The fires were unlit, and it seemed for a long time they would never happen again. But they did.

They just took another form.

How We Walk Through the Fire

In 2022, we took an ember from a canceled Dark Mountain spring gathering and blew on it. Its name, "How We Walk through the Fire," would become a series of workshops revolving around the eight fires of the ancestral solar year. These would be hosted online but involve encounters in real life. The idea was simple: we would build a cultural practice together that could weather the storm of converging crises using the "stone clock" of the equinoxes and solstices and four stations of the growing year, known by their Celtic names—Imbolc, Beltane (or May Day), Lughnasa, and Samhain. Each "fire" would be a marking, a celebration of our local wild and feral places, and a way of reconnecting ourselves with our bodies and imaginations and one another. After the first introductory meeting, we would collect sticks on a walk into our territory and build a small fire to mark the shift of season. For the second meeting we would bring back a fire stick and relate what happened when we held those fires: how we made them or failed to make them, what occurred in the wind and rain and snow, which creatures came, the glowing shapes the embers made. And then, as if gathering round a real fire, we would "show and tell" a piece of work, created from our encounters.

The workshops set out to host a regenerative culture that could navigate uncertainty and lay the tracks for a more "biospheric" relationship with the planet, a practice that was both modern and ancestral. Each of the fires explored different approaches, but all aimed to foster resilience and to strengthen our creative voices within an ensemble. They also provided a container for the rigorous inner work of relinquishment and restoration: what the poet Gary Snyder once called "hard yoga for planet Earth." At each of the eight "doors" of the year, we "walked through the fire," letting go of what no longer served and discovering what might bring repair and regeneration to a world—and a culture—in crisis. Above all, we voiced and celebrated what we experienced.

The shape of these fire-based meetings took the shapes of a ceilidh and a kiva, which is to say that aboveground, they mirrored the social gatherings of Scottish and Irish tradition: the convivial sharing of songs, poetry, and stories. But below these festive convergences a deeper meeting was going on, in what we called the "kiva" attention—named after the underground ceremonial chambers of the Southwestern Pueblo people where spiritual and cultural meetings are held, linked with the cycle of the year. To paraphrase the storyteller and psychologist Clarissa Pinkola Estes: this was "the fire beneath the fire," the bones of the mythos that underpins the narrative body of any transformative story. So although our meetings were creative and celebratory, there was an acknowledgment that their meaning was being played out in another dimension: we were making moves in our collective and inner lives, aligning ourselves with the movements of the sun and Earth.

It was this double-banded attention that the rowdy and edgy Lewes bonfires did not contain, in spite of their lively flames and burning tar barrels, or indeed any spectacle will lack which requires us to be the audience rather than the players. These ceilidh-kiva fires were not there to entertain anyone: they keyed us directly into the fabric of the world.

Into the Lexicon of Earth

Something happens when you sit outside by a fire with other people that would never happen if you were sitting inside a room or simply in the dark. Our kinetic, physical, feeling bodies, repressed by industrial mechanical living, break free; we tap into our bones, into our deep memory. You enter into the lexicon of Earth, a multistranded dimension, made of sound and shapes, position and feeling, that our senses immediately recognize. This "dreaming" language is hard to access without some kind of tool or container. Most of the time in our ordinary lives, we are trapped in what has been called left-hemisphere attention: a focus on looking at the world from a distance, controlling and editing what we observe, categorizing it in terms of data and numbers and "facts." To get to the feeling, immersive, all-connecting, right-hemisphere attention requires a bridge. You can debate and describe the world for a long time using words but never find a way to speak with it. A fire cuts through the blocking vocabulary of the clever rational mind and puts you directly in touch with life. You sit in the darkness with your fellows, human and nonhuman, and feel at home.

One of the hardest liberating moves to make, trapped as we are in the labyrinth of modernity, is to connect with the deeper ancestral part of our beings. No matter how well-meaning or smart we might be, this cannot be done in your head, sitting in a room or in front of a machine. You have to go out. You have to have a task in mind. You sit, you wait, you come back. You tap into the creative fire within your being, into the place "where memory meets imagination in the dark," as the great nature metaphysician Annie Dillard once wrote, and turn what you have experienced into a physical or creative form.[1] Then you share it.

The encounter is key. In a prototype workshop I held between lockdowns, one woman related how she left the task to the very last minute: after a terrible row with her family in the evening, she stormed out into the howling night and drove up into the downs above the town and stumbled toward a stone barrow that was way off the track. Somehow the storm inside of her became the storm outside of her, and she sat down by the stones, in the wild and the wet and the dark, and she laughed. I felt the ancient spirit of the place all around me, she said, and felt totally liberated in that moment.

We are held mute in our civilized cocoons and need a force from the outside to break us out of our restriction, germinating the seeds of the future we hold inside of us. For some, our husks are broken open by storm, some by a gentle persistent rain, or water. But some of us are cracked open by fire, by a big fire that comes unexpectedly and scorches us awake. However the germination comes, it feels like the end of something old and a beginning of something new.

In our Eight Fires series, held on two weekends, those breaking moments were shared in our stories in the second of the two-hour sessions. Everyone said no matter how they celebrated them—by themselves or with invited company—they felt emboldened knowing that there were others doing the same ritual in the wilds of Alaska and Finnmark; on the riverbanks and rooftops of Berlin, Vienna, Paris, London, Chicago; on the edge of an Indian forest or Hebridean coastline; in the Indigenous lands of Australia, New Zealand, and California. They brought their charred sticks and told the stories afterward of what had happened as they went through the fire. Of the creatures or trees or rivers that had accompanied them. Everyone was listened to and witnessed.

How do we "stay with the trouble" if we do not have a container that will hold us? If our feet are not on the earth? If there is not a ceremony that can bring us together in our endeavor to get to a future we might want to live in, no matter how challenging?

How do we connect with each other when we live continents apart?

What we discovered was that our frame of ecological and social breakdown, rather than turning people away, drew us out and together. As we reported back from the fires we had held in deep winter or high summer, we shared stories about the birds we encountered, the waters we honored, the winds we named, the plants we connected with, and the mythic layers of the earth beneath our feet. It felt, despite the current state of collapse, that this shared attention to place and time helped us engage in what the Aboriginal academic Tyson Yunkaporta describes as an "increase ceremony": to increase not the reach of a culture but its mycelial strength, to thicken the web of correspondences between place and people, ancestors, and the more-than-human world. An invisible network that you could feel in the online sessions even though we were not meeting in the physical world.

The Eight Fires

The fires are held at the eight points of the "stone clock," which are the ancestral points of the year that mark the seasonal shifts: the four cardinal points of the equinoxes and solstices. These Earth and sun calendars have been observed throughout the world for thousands of years, and in Britain they take the shape of two crosses, the cross of the sun (+) and the cross of the Earth (×), and much of our island folk dancing reflects this geometry interweaving these crosses in different combinations, from Maypole dancing to Scottish reels. In the Americas, these circles map the eight directions in space and are used as guides and medicine wheels for the changes human beings need to keep themselves in right balance. In Britain they are connected with time.

Being in time and space aligned with Earth and the sun are radical acts in a global consumer culture that does everything in its power to keep people deracinated and hostage to the drive of the 24/7 clock, forever caught on a whirligig of deadlines and desperation. This practice is a way to resist those forces, to connect with our ancestral knowledge and put us in sync with the Earth's rhythms and metabolism. The practice breaks us out of the hyperdrive of modernity, brings space and time within the mindset we are forced to live in, and liberates our imagination. It takes us out of the thinking (timeless) mind into the body and the heart (which holds the tempo of Earth); it takes us into deep time to connect us with our human and nonhuman lineages and the elemental building blocks of earthly life. And although this fire keeping can be practiced solo, it is interconnected with a mycelial network, tapping into the thousands of people who are holding fires and doing this work at the same time.

Each door marks a shift, a "technology," if you like, that allows whoever takes part to tune into these moments as a way to transform ourselves, following a road map that is ancient and global and has been practiced for millennia. As modern industrial people, we may be untutored in the way of Indigenous or archaic people, but that does not mean we are not able to do the work, wherever we live, or whatever our age or circumstances. We still live in the same bone houses, with an intelligence that can communicate with a sentient planet in its many languages, so long as we are willing to uncolonize our imaginations and open to the inner and creative tasks that lie ahead.

One of the shapes we have been working with in the series is the figure of eight, or lemniscate, which goes out into the world and returns with riches. It is the basis of many of the underworld myths we told during this year of fire practice, the shape the honeybee dances as she signals to her sisters where nectar and pollen can be found and brought back to the storerooms of the hive. Engaging in these feedback loops means we do not, as the ethnobotanist Frank Cook once warned us, become "end users." This is not only the shape that informs our encounters (going out into our territories and returning with testimony and medicine); it also is the shape we remember when we attend the fire. We feed the fire with the dead wood of ourselves, parts of our beings that no longer serve; we bring nourishment and sweetness to the collective with our songs and art and stories. These loops bring regeneration to the damaged fabric of Earth; create right relationship with lands that have been grievously harmed by centuries of extraction and exploitation; provide the set and setting for becoming a different kind of people.

Why does attending to fire matter? Because the ceremonial fire is a microcosm of the sun in the solar system, and also in ourselves, our core beings that act like the sun in our bodies, radiating outward and warming all who come near and gather. The act of making and attending to it connects worlds. When we align with our fiery star at these times, we get back into balance, in time and place. Our modern urban societies raise us to be cold, competitive, and mean but when we gather by these fires, our natural conviviality and generosity is brought out. The fire is a focal point, reminding us not only of the light that lives in the dark but also of warmth. In that radiating warmth we open like flowers; we are no longer hostile.

That's what I first noticed at that initial Uncivilization Festival when I sat down by the storytelling fire under the summer canopy of stars. I had been struggling, working as a writer and editor in grassroots groups for years, trying to convince strangers to decarbonize their lifestyles, trying to fit into communities where I was an outsider. Suddenly, I was among fellow writers and explorers, lovers of myth, creativity, and Earth, in company, at home. That was a good feeling. I did not go back.

Of course, there is not always a convivial fire when times are hard. "Staying with the trouble" is not just a phrase we might use in passing; it is a work, a task we do, an undertaking of inner alchemy and reconnection. The fires come as markers in time, so we can funnel whatever is going on in the micro and macro narratives of our lives into a hermetic space where things can be revealed, reworked, and acted on by the light of the sun and the dying and birthing cycles of the living Earth. Where we can undergo the tempering of our souls and spirits that makes us into real human beings, not just consumers of resources. Where we can come awake to and restore the damage done to ourselves, our fellow human beings, creatures, plants, and the places we live. We can do this knowing that going through this fire is not a one-off moment but a lifelong practice, a process, a decision, and that we are not alone in our task; that sometimes this passage is very hard indeed in ways that are difficult to articulate, as the rigid and unkind structures of civilization are broken up within us. Endurance and courage are needed, but the honing of creative work and sharing it, listening to and being witnessed by our companions, and the lightness we feel having gone through the fire, will anchor and connect us. The practice brings a depth and a resonance into our ordinary lives that cannot easily be described in words.

Looking at the converging social and ecological crises we face on the planet, at the rising carbon in the atmosphere, at the karma of centuries we carry inside us, a collective cry goes up by all those who feel powerless to counter its trajectory: *What can we do?*

We can change the dance. We can remember. We can relinquish. We can start again.

We can make fires.

The Sessions

In our series of workshops, we focused on different aspects of this regenerative practice.[2] Our discussions and teachings revolved around six main areas, with the technology of the practice and reflection on the process at the winter and summer solstices. The subjects were as follows:

Halcyon Days (Winter Solstice, December 21). Our first fire gathering revolved around the fourteen days that surround the winter solstice, when it was once believed that the mythic Alcyone, transformed into a kingfisher, would nest by the shore in peace because her father Aeolus, the god of the winds, had calmed the waves. This practice offers an invitation to pause at the turning of the year, to enter a contemplative space and to create a piece of work from within it. The series began in the dark transformational moment of solstice, following the track of worldwide archaic and Indigenous cultures that have always looped back to their ancestral beginnings in order to know how to proceed towards the future.

Kinship with the Beasts (Imbolc, February 1–2). This collaborative journey started by looking at our core human relationship with the animal kingdoms. Imbolc traditionally marks the first stirrings of spring in the Northern Hemisphere, and as amphibians begin to move and the plant world unfurls, we asked what it means to reenter kinship with our fellow creatures and find ways to articulate and to strengthen that archaic relationship in this time of emergence.

Walking into the Wind (Spring Equinox, March 20). As the year shifted, we stepped out and connected with the elements, and especially the wild wind, in times of storm and climate breakdown. By tuning into the weather systems of our local territories, we examined the art and practice of liminal walking as we crossed the bridge from the dark watery realms of winter into the light and air of spring.

Plant Dialogues (May Day/Beltane, May 1, and Summer Solstice, June 21). These two celebratory fires heralded an immersive voyage into the growing world of plants to discover how we might reentangle ourselves with its intelligence and beauty, working with the key leaves, flowers, and trees of spring and midsummer. We experienced how plants help root us in place and time and remember the role human imagination plays in communication with the sentience of the planet.

Waterland (Lughnasadh, August 1–2). This time of harvest is often celebrated by visits to springs and holy wells to give thanks for water's powers of restoration. As the summer's heat intensified, we focused on the joy of our physical connection to water and how it runs through our lives in our language, our myths, our stories, our poetry, and our dreams. We explored bodies of water local to us—lakes, rivers, streams, and seas—in a time of ecological crisis.

Mythos and Mycelium (Autumn Equinox, September 23). An exploration of what it means to make art that engages with myth and the underground networks of Earth. Myths threaded themselves like a mycelium through the series, from the shapeshifting Suibhne (Imbolc) to Olwen of the White Track (summer solstice). As the year shifts toward the underworld, we went in search of the myths and stories held in our own local territories and ancestral memory and the mysterious role fungi play in the fabric of Earth.

Honouring the Ancestors (Samhain, October 31). At the hinge of the ceremonial year, as the light descends toward the dark months of winter, we looked at how to maintain a dramaturgical practice. Our yearlong journey deliberately engaged with a time frame that goes beyond this civilization's history, toward a past that is held in the planet's rocks and its many life-forms, as well as in our own human "bone knowledge." In this time of unraveling, we turned to our ancestors who can teach the creative moves we need to make in our everyday lives.

Gathering Sticks for a Fire

This practice of gathering sticks can be used as a basis for any of the fires, adjusting the questions to the seasonal cycle. For example, for gathering sticks for a fire at Samhain, find a time to walk into your territory, bearing the time of Samhain in mind, taking note of the shift of season, the change of mood and temperature, the scent of the wind, the color of falling leaves, the sounds of the insects around the last ivy flowers, birds gathering in the skies at twilight.

Make time to lie on the earth and look up at the sky, and to notice the clouds and the change of light. Tap into how this makes you feel. On your way home, gather some sticks for your fire. Make sure the twigs or small branches are dry (unless they are ash, which burns when it is green).

As dusk comes at Samhain, you can make a fire using your bundle of twigs as kindling, either in the territory itself or at home. This can be as small or large as you like and will depend on where you live, but even if you don't have a place available, a small twig fire in a can or metal dish will do just fine!

Once lit, whether a solo or convivial fire, sit with the fire and tend to it for at least fifteen minutes to tune in to this moment of the turning solar year. Engaging in the eight "doors" of this year is an imaginative practice to help reflect on our life in sync with the season and the times we live in, to rekindle the ancestral knowledge we hold between us and our relationship with the sun.

In the growing, birth, and death cycle of the year, Samhain marks the energetic shifts from the autumn harvest toward the underground roots of winter. As the leaves wither and fall, the composting of the summer's growth begins, a process that feeds back to life on all levels. As the months darken and we enter into the spirit of winter, it is also a moment of treasuring what matters, of contacting the deeper parts of ourselves, the bone knowledge we can trust to help us weather the challenging times ahead, not just for ourselves but for the Earth we are all part of.

As you watch the fire, you might like to consider or speak out loud what you are letting go of and what you are stepping into.

You might ask: What can I relinquish that will feed the fire, the spirit of the world? What is being required of me in these times of downshift? Who are the ancestors who can help us remember who we really are? What does the fire tell me?

Make sure the fire is well out before you go indoors.

At daybreak the next day, go out and greet the sun as it rises.

notes

1. Annie Dillard, *The Writing Life* (New York: Harper & Row, 1989), 26.
2. A creative workbook on the Eight Fires workshop was published in *Dark Mountain—Eight Fires*, no. 24 (October 15, 2023).

Firebird

Em Strang

I

Once upon a time there was a double miracle.
Nobody believed it but everyone knew it to be true.

It took place on a Friday or perhaps every Friday,
as the great bird appeared in a burst of flame.

The wings were tremendous curtains of fire
and the throat was a flute made only for song.

We knew the inferno would be quick and happy,
since this is how freedom tends to be.

We watched as the tail went up before us,
this fabulous bird incandescent with faith.

A quiet heaven fell as a small hill of ash,
the tilth of a being who knew how to die.

Those of us sleeping gave birth to the doubt—
that horrible weight in the twist of the gut.

But the fledgling arose—we watched it together—
from the ash of forgiveness and the embers of grace.

II

And then I realised that I'm the fire.
But the fire—is not me.
Although at the same time, I'm the fire.
—paraphrase, Daniil Kharms

Come into this (im)perfect timber forest
to look at the (un)broken man, the one

who's put the fire out
or thinks he has.

Watch as he stamps the pit
in his black boots, cursory,

spitting at the embers,
his heart nested in the ash,

waiting
(as a fledgling swallow waits).

The sun throbs inside his head,
a thick dunt like a drunken fist.

For forty miles in all directions
the trees too are waiting

and the river is
a lisp between boulders.

Even if he prayed for rain—he won't—
it wouldn't fall.
Watch how the fire comes
to unzip the man—

a conflagration
in which he receives himself:

bull, raging butterfly,

double miracle
of flesh and spirit.

Watch him begin again,
the bird in his chest screeching.

III

After George Seferis

That whole night we were full of misery,
my God, how full of misery!
In the end, cavernous hearts, black guts, the trauma of ages
pumped into our sluggish blood.
Men with cocks as cruel as crowbars
broke down the women and broke down the children,
splitting them in half, taking them away from themselves forever.
Nobody heard us crying, calling.
The moon too was cut in half and the sun never rose.
The whole night and the whole morning
we were full of misery,
all light kept in shadow by the murderers.
Do you remember their laughter—how full of death!
Then the sun came, or something like the sun,
a vast explosion of light and heat
in the shape of a bird beyond our imagining.
We stopped crying, stopped calling
and lay down on the waiting earth.
The bird shone all of creation into being
before the men could stop it, after the men
had wept. *Inexplicable,* you said, *inexplicable.*
I don't understand people:
no matter how much they drown in darkness
they hold this unquenchable fire in their hearts.

The Star We Know

Isaac Yuen

Two rather important things happened 170,000 years ago.

The first was that a beam of light, having engineered a convoluted escape from the core of our favorite sun, flew free for eight and a half minutes through space before falling under the spell of yet another yellow star—this time of a lily variety in the *Hypoxis* genus, growing by the Lebombo Mountains in South Africa.[1]

The second was that back in the hottest, densest part of the sun, two hydrogen atoms were in the midst of completing their merger to form a new helium venture, which usually results in a tiny bit of shed mass and a rather large burst of deadly gamma rays.

What commonplace events! you might think—with your big brain privy to the fact that photons shower down on Earth all the time, or that fusion events occur inside the heart of every G-type star. But significance can be relative: sometimes it is realized only in hindsight, after 170,000 years, like when archeologists unearthed evidence of former *Homo sapiens* roasting vegetables in the form of a certain lily tuber growing by the Lebombo Mountains in South Africa. Sometimes such a lengthy delay in recognition is forgivable, especially when it comes to light that the light that makes it possible for you to read this fact also took around 170,000 years—plus eight and a half minutes—to escape the sun's core to reach those receptive retinas of yours, mellowing out over that span of time into a less lethal and more eye-friendly form of radiation.[2]

Thus is relevance revealed; without that certain incident of former photosynthesis, present day you might not be around, there being a lack of easy calories to fuel the evolution of big brains and deep thoughts. And without this past episode of vintage sunshine, present day you might still be groping in the dark, there being no light to spark curious minds to read on about that giant ball of fire ninety-three million miles away, that chief purveyor of miracles happening every second of every moment on this little rock, our motley Earth.

But wait! you might say—being a night owl of the figurative variety and not of the literal great-horned variety, whose call would most likely begin with *who*. For you are not reading this by starshine but by lantern light or candle flame or the constant pixel gleam of the latest modern marvel. But separation can be illusory: these seeming disparate forms of incandescence are in reality part of the same conversation, not wholly original in their ancestries. Take the lantern. Contained within each drop of kerosene is the collective vitality of expired phytoplankton, most likely from the age of the dinosaurs, having soaked up rays on balmy seas before sinking down to be consolidated into crude. That warming glow and caramel scent emanating from your artisanal beeswax candle? The result of a long chain ending at the tips of wax glands mounted on bees fueled by glucose derived from honey converted from nectar collected from flowers, also known as the public fronts of miniature solar power plants. Even the wind and water that spun the turbines that created the current that rode the power lines and fed the sockets to charge your smart screens are primarily sun driven, caused by an uneven tanning of our spinning, tilted globe, closing in and veering out in a perpetual orbital number. Perhaps it is better to conceive of these energy permutations as dialects of the solar mother tongue, translations in nuclear or electromagnetic, exchanges carried between chemical to mechanical or potential to kinetic. Even should some essence or nuance be invariably lost along the way—chiefly through increments of dissipated heat—there is no need for concern. The sun is there to make up for the difference with sheer abundance, promising the world to the smallest diatom and the largest sequoia, kindling the laziest, lounging iguana to stir (eventually) into action, favoring the continuation of its blue Earth with yellow lilies over the relegation of it to the status of infinite hiatus, that bleakest and most unchanging of fates.

There are those who do not subscribe to solar hegemony. There are those who are wary of relying on something so remote, so magnificent, so blindingly benevolent. Most of these mistrustables are bacteria, coalescing into colonies around cold seeps and hot springs, clustering by hydrothermal vents along midocean ridges, shunning the light to become adherents to an alternate life philosophy. *No to relying on radiation streaming down from on high! Yes to oxidizing inorganic compounds of one's own choosing!*

But while chemosynthesis appears to have freed its practitioners from the whims of a distant star, most of them remain linked in some fashion to the sun-based network. Many a vent-dwelling bacterium digesting their black-metal sulfide meals from below still require oxygen generated from the world of green above, and many of the invertebrate neighbors they depend on for rides to secure new vents and new opportunities still derive their living from photosynthetic ventures. There is no escaping the fact that almost everything in this world is depended on and depended on by degrees, save perhaps for certain groups of single-celled recluses, and even then there is connection and relation through old history, back to the time of the Great Oxygenation Event, when these anaerobic survivors were exiled to the darkest recesses of the earth. Even now, they work to cultivate their being in the strict absence of oxygen and sunlight, toiling in the depths of sediment and landfills and intestinal tracts, perhaps pining for the day when the world is utterly transformed once again.

Indeed there will be a day when the sun sets, and not the daily, fleeting version—which, strictly speaking, is not something that happens, at least from the sun's point of view. The last, true sunset will happen long after the first, great sun swell, when the star will turn to helium to fuel its fires after depleting its present hydrogen supply. Five billion years from now, the sun will expand into a red giant, engulfing Mercury and Venus and a desiccated Earth before giving Mars a good scorching. The last, true sunset will happen long after this official retirement from the main sequence stellar circuit, which is predicted to proceed without any supernova fanfare, for G-type stars are not showy enough to broadcast their legacy across the universe with enough dazzle to outshine galaxies. No, our Sun, being modest, will simply shed its mantle of hot, leftover plasma and leave behind a nebula, brief but tasteful, before downsizing into an Earth-sized white dwarf. Thus will begin the longest but loneliest phase of its life, as an empty nester diminishing across the eons, having long forgotten the pitter-patter and jibber-jabber and flutter-flapping of baby capybaras and teenage howler monkeys and Canada geese flocking home, going on without companionship save the dust of its former creations, remnants and reminders of a time long since passed.

But maybe the last sunset that truly matters need not happen the moment the sun goes dark. Perhaps the last sunset that truly matters will simply be one seen by those still capable of becoming spellbound at the sight of a fiery disc slipping beneath the horizon, by those still keen on seeing the transition as a moment for reflection and stocktaking, by those still able to be moved by the fact that there is always a final instance of everything, whether it be for a being or for a star, regardless of point of view. By those of us, grateful to have been touched by last light.

Steeped in such heavy melancholy, we may presently look up to the heavens, wishing and searching for better-fated earths, circling better, flashier suns. This is only natural, and in some ways healthy, to ponder proactively, astronomically, with a dash of humility. How Terra-centric and Sol-devoted we must be to assume that here and only here is best for hosting paradoxical frogs and chicken turtles and striped pyjama squids, especially when we have scarcely begun to inventory the skies! What if exoplanet KOI 7711.01 proves a goldfish's paradise, dotted over with giant lily pads spangled across perfectly circular ponds? What if one of the twin worlds orbiting Kepler-371 proves ideal for ibexes, being filled with suitable inclines without any hazardous felines?

To that effect, astrophysicists have begun conducting preliminary surveys, highlighting two dozen systems out of four thousand that may be "superhabitable."[3] Some contain stars that burn cooler and longer, allowing life tens of billions of years to evolve in leisure. Some feature planets that are denser and weightier, allowing for stronger magnetic shields and less-flighty atmospheres. Perhaps one of these locations will ultimately prove more suitable for the orange-spotted filefish or the Namibian quiver tree, both of whom are struggling mightily with the weirding weather of our present climes. Yet as attractive as such off-Earth options may be, most homegrown creatures appear reluctant to relocate. This may partly be due to logistics. Furniture-shipping costs from here to the constellation of Lyra are borderline criminal, and the light years to and from the nearest superhabitable planet can neither be considered carpoolable or migrateable. But perhaps for most, the hesitancy to begin anew comes down to the matter of attachment, of being too invested in existing connections to entertain the prospects of acclimatizing to new gravities and day lengths, even should the extraterrestrial grasses be confirmed at a later date to be greener on those other sides.

For aliens from super worlds visiting us for the first time, our living arrangement around an unremarkable star on a backwater arm of the Milky Way may seem inexplicable. Maybe we will never be able to convey the mix of flaws and charms that makes this setup so delightful and unique, on our little rock, around our yellow sun. Or maybe they will realize they can understand this je ne sais quoi only if they too embrace the tangle of life and light that has come to nourish each and every earthling from seed to soul to dust, that ineffable quality that has convinced us along with the basking sharks and the sunbeam snakes and the yellow archangels in bloom to remain earthbound and sun-wise despite all the turmoil and struggles, and despite the knowledge that there may appear to be many a better where.

notes

1. Lyn Wadley, Lucinda Backwell, Francesco D'Errico, and Christine Sievers, "Cooked Starchy Rhizomes in Africa 170 Thousand Years Ago," *Science* 367, no. 6473 (2020): 87–91, https://www.science.org/doi/10.1126/science.aaz5926.
2. National Aeronautics and Space Administration, *Our Star—The Sun Lithograph*, July 2013, https://www.nasa.gov/sites/default/files/files/Sun_Lithograph.pdf.
3. Dirk Schulze-Makuch, René Heller, and Edward Guinan, "In Search for a Planet Better Than Earth: Top Contenders for a Superhabitable World," *Astrobiology* 20, no. 12 (2020): https://doi.org/10.1089/ast.2019.2161.

Buried Sunshine

Catharina Coenen

I don't remember when my Oma Lotte's best friend, Tante Emmi, came to visit and brought with her a copy of the *Westdeutsche Allgemeine Zeitung*, the most prominent newspaper in the coal-mining cities of Germany's Ruhr region. I might have been ten. Or perhaps thirteen? Which means the *WAZ* that Emmi brought was from sometime in the late 1970s, the beginning of what we now call the era of deindustrialization, the years of recurrent mass layoffs and labor strikes. But the pages of Emmi's edition of the *WAZ* did not show miners marching in grim solidarity. Instead, they sported grainy photographs of coal-blackened rocks laced with imprints of filigreed ferns, stenciled pine cones, scaly bark. I remember Lotte's and Emmi's excitement, their insistence that these pictures were significant, a marvel, mind-blowing, the kind of thing I ought to learn about in school.

I took the pages, thanked them, read, but struggled to match their exhilaration to inky blobs on coarse newsprint. What did I know of leaves? Of coal? Only that leaves grew on trees. Only that coal came from the ground. Only that, when I was small, my grandfather had shown me how to shovel coke into the furnace in our basement before the black, loose pile of this hot-burning coal derivative was replaced by tanks for heating oil.

I stacked Emmi's *WAZ* with other pieces of paper I knew not what to do with or how to throw away.

In seventh grade, my teacher drew a yellow sun on the blackboard, then a green plant. One of the sun's yellow-dashed rays touched a leaf.

"Photosynthesis," he said, "is how plants catch the sun's energy."

He drew a yellow star on the green leaf to signify that the sunbeam was now, somehow, part of the plant. Next, a brown bunny appeared by the leaf, whiskers twitching, ears perked.

"When animals eat plants, the sun's energy is transferred from plants to animals," the teacher said.

He switched back to yellow chalk and drew another star, this time inside the bunny's belly: "All energy we need comes from the sun."

It was a magical concept, this bright body in the sky filling forests, fields, and rabbits with little yellow stars.

By the time I started studying biology at a university, the sun in the bunny's belly had become depressing, frightening. Like other animals, I am a heterotroph. To live, I must eat. To eat, I must take life from other animals or from plants. There is no escape from this accumulating responsibility for deaths that drive my life. Each lettuce leaf, each grain of wheat is full of living cells.

I studied biology because I loved them: the sun, the bunny, and the plant. Their shapes, their colors, the sun's fire that drives plants and animals to grow, to live, to love. Yet in the concrete towers that were my northern German university, I learned in ever-greater detail that to be human means to kill. As long as I am alive, I will swallow them, then breathe them out: the bodies of lettuce, rabbit, wheat, their cells disassembled and burned inside the biochemical furnaces of my digestive system and my mitochondria.

In a country that had burned millions of human bodies into smoke and dust, what I wanted, more than anything, was a blameless life. Unlike my parents, I had been taught in school not only about sun and bunnies but also that my grandparents' generation chose to participate in genocide, to benefit from it, or, at the very least, to look away. What no one taught me: how to live with seeing, with knowing, that what I want or need takes lives.

So there I was, alone with a diffuse, all-pervasive guilt that made getting out of bed, eating, and taking the bus to university feel like lifting bricks of lead. There I was, twenty-one, and twenty-two, and twenty-three, a privileged student at a university, studying what I loved. What should have energized me—the desire for a bright future, a profession that would fill me with fascination and joy—seemed to crumble and collapse under the weight of bodies hewn each time I wanted something, took a breath.

Emmi, Lotte, and my mother did not go to university. They all grew up in the industrial area adjacent to the Ruhr River, colloquially known as Der Kohlenpott, "the coal pot." Sometimes the word *Pott* is thought to refer to the fact that the Ruhr flows through a pot-shaped valley. Other times the "pot" is pictured as a "melting pot"—the area where workers with different German dialects melded languages with Irishmen, Frenchmen, and Poles, amalgamating cultures, fortunes, and histories. For two centuries, people kept arriving—from the south and the east of Europe, from Turkey, from Africa, from South Korea—not because life was easy in the Pott (it wasn't) but because it was possible. Where miners came from, hunger killed. The Pott turned everyone into a *Kumpel*, a word that simultaneously signifies "miner" and "mate," the camaraderie of having each other's backs while working dangerously far below the ground.

Everyone in the Pott lived on money paid by mines, whether or not they toiled *unter Tage*, literally "below the day." Both of my mother's grandfathers worked in tool factories that shaped coal-fired steel. Her father, Alfred, was one of thousands of unemployed young men during the years of economic depression before the takeover of the Nazi regime. Along the Ruhr, just as elsewhere in Germany, workers starved even if they had a job. Children died. Inflation shot up to a thousand percent per year. Hunger and deprivation drove political extremism, racism, and violent unrest. The Nazi Party rose on promises to end chaos and hunger by "opening" Eastern Europe for German agriculture.

After the war, with shrapnel buried in his shoulder, Alfred found an office job in a brickyard in Waltrop, where he befriended Emmi's husband, Theo. Theo was *Oberbrandmeister*, the "superior fire chief," not a miner of coal but an expert in its use. The year was 1948, the Ruhr area British occupied. The Deutsche Mark had just been introduced as a new currency in the western occupation zones to end an economy solely run by rationing. In Waltrop, as in brickyards all across Germany, coal fires burned to harden clay dug from the ground. An estimated 160 billion bricks were needed to rebuild German cities after the devastation of the bombs. Nearly eight million Germans had died during the war, either as civilians in the fires of bombs dropped from the sky or as soldiers at the front. Another two million former soldiers were still in captivity. Finding knowledgeable workers, like Theo, to run the brick ovens was no small feat.

Most of the bricks that rebuilt German towns came from ring kilns, like the one Theo ran in Waltrop. In these kilns, fire traveled through fourteen to twenty chambers, arranged in a circle or an oval, into which bricks were continuously unloaded and reloaded, day and night. Inside the chambers, each raw brick was stacked at a specific place and angle, leaving regular open shafts below holes in the ceiling through which ground coal would be dropped into the fire. The timing and amount of ground coal added from above took knowledge, precision, and experience to heat and cool the bricks at the right rate. This was Theo's job.

Bricks from coal-fired ring kilns were even, dependable, strong. Coal had built the iron smelters and factories along the Ruhr that first made Hitler's planes and tanks, then drew the Allied bombs. Now more coal was needed to rebuild not only what fire from the sky had destroyed but also livelihoods, an economy, a country, a democracy.

There was no time to think about what coal was made of, where it came from, or what it meant to burn and send its smoky carbon back into the sky.

My mother never felt at home in the mining culture of the Ruhr. To marry my father, she moved an hour away to a town that had sprung from salt and wheat and trade. Where I grew up, coal arrived by train and truck. I never saw a miner or a mine until my university program demanded that each student participate in a study tour to learn where biologists could work.

After visits to chemical factories and government laboratories along the Rhine River, where dances of fish and water fleas in river water were used to monitor toxic leaks from factories, sewage treatment plants, and mines, we visited one of the few remaining coal mines along the Ruhr. Our guide instructed the female students to watch a documentary while the male students toured the mine belowground.

"Miners think women *unter Tage* are bad luck," one of my classmates stage whispered as she rolled her eyes.

We protested. The guide explained that there weren't any showers for women. Our male fellow students said they would not go unless the women also came. The guide conferred with his superiors, who said that the women could use the supervisors' shower room. And so we changed out of every stitch of clothing to don white coveralls. Not even civilian underwear is allowed *vor Ort*—literally "at the place," which is miner's speak for where coal is broken from the walls and coal dust fills the air. Any spark from synthetic fabrics could lead to devastating explosions of methane gas.

As the elevator dropped at a disconcerting twenty-five feet per second, daylight disappeared. I don't remember how far down we went—the lowest horizons of coal mines in the Pott extend more than three thousand feet below the surface, which means our precipitous ride might have lasted two or three minutes. To reach the *Flöz,* the coal seam, we took an underground train, then clambered into a low black *Streb,* where steel shields to our right pressed up against the rock ceiling, preventing it from crushing workers crawling through Earth's past.

A three-foot-thick coal seam is about five thousand years of compressed, tropical swampland forest from the Carboniferous. Five thousand years of scale trees, tree ferns, and mosses breathing in carbon dioxide and "fixing" it into leaf and wood and spore. Five thousand years of sunshine in wet, tropical latitudes, before plants invented flowers, before the land that will become the "pot" is shifted north by continental drift.

Like any colonizers, miners named underground forests to honor what they valued at the time: Margarethe, Katharina—saints, empresses, and queens (and also the names of their own daughters)—then, later, numbers to reference an industrial rationality. They lacked the perspective to name the coal seams for what they were: bodies of ancient ferns that grew like trees, of towering horsetails, of club mosses that scraped the clouds, of thirty-inch-long dragonflies, two-foot scorpions, and nine-foot millipedes, of yard-thick layers of pollen and of spores released in enormous masses to ensure the forests' future.

We turned on our headlamps to see coal glitter as the plow broke it from the wall, dropping it onto a conveyor belt. Everything was hot, everything was loud. Breathing—through the fast-blackening white cotton cloths we tied across our mouths and noses—was hard. And I understood: this, right here, the sweat, the dark, the dust, the thousands of tons of rock above our heads, was what was meant by "work."

Breaking coal, first with a pickax, later with jackhammers, then shoveling it into trolleys, was what made money for miners who were paid by the trolley load. Coal kept hunger at bay. *Kohle machen,* to "make coal," in German, is still synonymous for earning money. My ancestors ate, and lived, and gave me life because men hammered holes into the earth. And I emerged from dark and dust not with solutions, not with an idea about how to live with the mountain of guilt I had been carrying as long as I could think, but with a bone-deep appreciation for the bravery of men who dropped their bodies into darkness and danger, day after day.

Like most of my fellow students, I could not talk about what I learned at university at home. Worries about climate change seemed abstract, far-fetched to parents and grandparents who had survived so much hunger, so much war. When my only two friends at university both decided to move away, I noticed a poster for the Fulbright study-abroad program pinned to a wall outside the international office. I pushed the door handle down, pushed myself through the door, and asked the secretary behind the counter for the application forms.

Weeks later, I sat for an interview with three or four professors in a bleak classroom. I had no one to guide me to prepare for this, no one to help me guess what they might want to talk about.

"Why do you want to study plants?" one of them asked.

"Photosynthesis," I blurted.

Inside my head, I was back in the darkened microscopy lab. We'd learned how to digest away the cell walls of tobacco leaves, then centrifuge the tender, wall-less cells, gently maneuver them onto glass slides. Under the microscope, each cell was a translucent sphere, its membrane delicate as a soap bubble. Inside each bubble chloroplasts glowed like verdant continents, drifting in oceans of cytoplasm. In the microscope's beam, their green seemed the epitome of how to be alive, transforming radiance to growth: the only blameless way to live.

Across the classroom desks, I tried to describe the moment: "Plants don't need to eat anyone. They know how to take energy directly from the sun."

The professors looked at one another, silent. They shifted in their uncomfortable chairs.

"What about agriculture?" one of them finally asked.

I, too, wriggled on my plastic seat. I drew a breath.

"Yes," I said. "Agriculture. Sure."

And just like that, I was on my way to a US land-grant university.

Some people see Jesus or the Virgin Mary when they are close to death. Some see a dead relative or a long-lost love. But my grandfather, Alfred, emerging from a fog of morphine and pain, from fever dreams that put him at the Russian front, squeezed Lotte's hand and pointed: "Look, there's Theo standing over there. He has come to fetch me."

Theo. Not his mother. Not Jesus. Not a saint. Not a comrade from the war. Theo, maker of bricks that rebuilt his world. Theo, master fire keeper, shoveler of ground coal.

Emmi's coal-black pages of the *WAZ* resurfaced whenever I sifted through my stack of random papers, wondering what to toss. Each time, I'd slide the yellowing newsprint back into the "keep for now" pile, although I had no idea what I might need it for. The pictures of black leaves on rock were an enigma in the back of my mind, a riddle planted by my grandmother and her friend: the beginning of the end of mining in the cities of the Ruhr, the need for understanding passed from grandparents to child. The miners' labor, their exploitation of forests past, built my generation's economic security, buying me a youth away from child labor, work that regularly killed men, persistent hunger, the cold. Coal bought me time to read, to learn, to wonder, and to understand what my grandparents never knew: that they were building a safer, easier life for me at the expense of spending Earth's inheritance, the bodies of forests that had banked away the sun, that had inhaled vast amounts of atmospheric carbon and buried it deep in the earth.

My mother learned to crawl, to talk, to sing in Essen's Steeler Strasse 372, a gray block of rental flats. Wherever my grandmother pushed her in the baby carriage, wherever they walked, mines ran underfoot. They lived surrounded by *Kumpel*, by openings in the ground that breathed out methane gas, by blind shafts drilled to probe for further coal, far above men that picked and hammered their way forward through the compressed forests of the Carboniferous, the period 359 million to 299 million years ago, when the Pott was a vast tropical swamp, home to the first four-footed animals attempting life on land.

I grew up a short distance from where coal seams rose close enough to the surface to build the cities of the Ruhr. But I realized that the Ruhr's culture was not just my grandparents' but also my own only when I stumbled upon an online *Ruhrpott Lexikon*: *Appelkitsche, Bedrullje, Hümmeken, Pullefass*—here were the sounds of my childhood, conjuring the gnawed-clean core of an apple, serious trouble, a small paring knife, a bathtub. Only seeing this vocabulary recognized as its own "lexicon" helped me to understand these sounds as the melded language of coal-mining families, something my parents and grandparents had carried with them and imprinted on my brain as synonyms of "safe" and "warm" and "home."

The German word for coal mine, *Zeche*, comes from the Middle High German word for a series or an order. Initially it described people organizing themselves to run a mine. Eventually, the mines themselves became known as *Zechen*. Because the members of the group paid dues, *Zeche* also came to mean a bill for food and drink at a pub or restaurant. No one considered that each *Zeche*, each ton of coal removed and burned, opened an invoice to be paid by future generations of humans, animals, and plants.

A decade after my Fulbright grant first sent me to America, I had become a biology professor and also "Answer Aunt." Each time I landed at Stuttgart airport, my nephew was there, pressing small palms and nose and forehead against the glass that separates the international baggage claim from the arrivals hall. Minutes later, he would slip his sticky hand in mine, saying "I have *sooo* many questions!"

The summer before my nephew turned twelve, he had a question I was actually qualified to talk about.

"No, I want to know how it *really* works," he snapped at my sister, when she tried to explain photosynthesis by hand waving about sunshine, sugar, and green leaves.

And so we sat on my parents' rug in front of the bottom bookshelf—the cross-legged botanist, the almost-twelve-year-old, and my parents' leather-bound thirty volumes of Bertelsmann's hyperillustrated encyclopedia, bought decades earlier to satisfy my curiosity.

This is an atom. See the electrons? Here's how atoms join into molecules. Here's where this particular electron likes to hang out, zooming around its special orbital in chlorophyll, and here's how a light particle, called a photon, kicks the electron into a different orbital. From which the excited electron leaves the chlorophyll, passing from one electron carrier to the next, losing a smidge of that light-given punch with each pass. Here's where the electron meets another chlorophyll and gets another photon kick, and here's where a chemical picks it up, to hurry it over here, so it can, eventually, convince carbon dioxide to turn into carbohydrate. No, carbon dioxide does not turn into oxygen. The oxygen that plants breathe out comes from when the first chlorophyll rips off water's hydrogens, to replace the electron it lost, see? No, not easy. Yes, you do need chemistry.

For his twelfth birthday, I sent my nephew a molecule modeling kit. When I next came to visit, plasticky carbon dioxide, oxygen, and water floated above his pillow from invisible nylon strings, a home-made mobile of coal-black carbon, white hydrogen, and bright-red oxygen. To *grasp*: another word for "understand."

Over the past five years, summer droughts have damaged one-third of trees in Germany and have shrunk German rivers down to record lows, killing fish and exposing live grenades from World War II along the Rhine's widening gravel beds. In 1992, Germany signed the UN climate agreement. Since 1997, Germany's parliament has also listened to speeches from Holocaust survivors every year on January 27, the day Soviet soldiers liberated Auschwitz. In my mind, both parliamentary actions embody my generation's commitment to exposing, to understanding, to the sitting-with-distress we learned in school. Guilt is not something to put down or something we can solve alone. If we want to live, we must take life—but we also must refuse to look away from where our taking causes pain. If we want to survive together with the other glowing, crawling, singing species who radiated from the Carboniferous, humans must live within the budget of today's sunshine, water, wind, and the earth's own body heat.

On September 14, 2018, the last *Kumpel* broke coal for the final time, in Bottrop's Zeche Prosper Haniel, opened in 1856 and the last coal mine in the Pott to close. Less than a year later, the mining cities of the Ruhr region began a project they call Solarmetropole, the "solar metropolis." Solar cells use the sun to make electrons flow, just like plant cells do. And just like in photosynthesis, photovoltaic current can be used to split water into hydrogen and oxygen. Essen is now headquarters to ThyssenKrupp, a group of old steel and chemical companies planning a "hydrogen hub" in Duisburg, where water splitting will supply both hydrogen and oxygen to steel industries through pipes that currently still carry natural gas. By 2050, this switch could cut carbon emissions of steel production by 95 percent. Solar energy and hydrogen regularly make news on pages of the *WAZ*.

But unlike photosynthesis, photovoltaics do not pull carbon dioxide from the atmosphere; they do not bank electrons in organic molecules. They make no wood, no grass, no coal. Storing solar energy remains a problem. Using the current to produce hydrogen is one solution, but one that lacks the efficiency of photosynthesis. In the energy economy of biology, we, the organisms, are one another's batteries. The sun in the bunny—and the radiance stored inside my liver, muscles, and brain—is food and energy for someone else. The bunny, the standing forests, prairies, bogs: they are, yes, carbon banks. But just like me, they also are themselves: sentient, alive, offspring and ancestors, links in a chain of sparks, of sunlight caught and passed along.

My parents probably tossed Emmi's *WAZ* into the recycling bin when they moved out of the house where I grew up. By that time, I was far away in Pennsylvania, in another town built on coal and steel and rust. But images of fern leaf, cone, and bark on blackened rock live on inside my mind. And with them live Emmi's and Lotte's voices, the excitement in their eyes: their astonishment at imprints of millions of years of sunshine feeding life on Earth.

Instructions for a Wampanoag Clambake

Lucille Lang Day

Wade into Popponesset Bay
to collect some Rock People—
old round stones
smoothed by the tide.

Remember Moshup, the giant
who predicted the arrival of white men.
When he said good-bye
to the People of the First Light,

he turned into a whale.
Find a place in forest shade,
make a circle, and dig
a shallow round hole for the stones.

Moshup's friend, the giant frog,
came to the cliffs and wept.
Changed into a rock, he still sits
at Gay Head today and looks out to sea.

Before finding dry wood for the fire—
your gift from the forest—
notice the shape of the hole
and the stones: All life is a circle.

When the tide is low, gather
quahog and sickissuog clams
and plenty of rockweed,
whose stipes are loaded with brine.

Light a fire over the stones
and when the Rock People start to glow,
pile rockweed on them.
This is their blanket.

As saltwater is released
from the stipes and steam rises,
add clams, lobsters, corn,
more armfuls of rockweed.

This is the *apponaug*: seafood cooking.
Now thank Kehtannit, who saw
the frog's sorrow and turned
him into a rock out of pity

and taught the People to use
the Earth, plants, animals
and water to care for themselves
after Moshup left.

The deer will always make you laugh
the mountain lion take your side,
the Star People shine on your path
if you do it this way.

Sky of Plankton, Ocean of Stars

Eiren Caffall

It is advisable to look from the tide pool to the stars, and then back to the tide pool again.

—John Steinbeck, *The Log from the Sea of Cortez*, 1951

I was twenty-two years old that July 1993 when the call came. I stood in my parents' kitchen talking to my father and the telephone rang and I answered it and there was the news.

My father's nephrologist had given me an ultrasound to rule out the genetic polycystic kidney disease (PKD) that was killing my father, had killed our family for two hundred years. A drowning illness, PKD fills the kidneys with cysts, each one a toxic tidepool of bacteria and fluid and chemicals. Dr. B didn't expect to find anything. I was too young.

"This is Dr. B." *It's Dr. B,* I mouthed to my father, who was sitting at the kitchen table. He nodded, his body rigid.

"I should have made an office appointment for you," Dr. B said, "but I thought you'd be fine."

I couldn't catch up to the sentence. I stared at the electric stove in the corner, the pottery jar full of kitchen tools, the antique wooden mallet, scarred and stained.

I walked to the picture window until the cord on the phone pulled taut, the limit of my distance from my father. I stared at the hill that the English settlers of Massachusetts called East Mountain. Beneath that ridge, the Wôbanakiak: Amiskwôlowôkoiak, "people of the beavertail hill," made their village. Ten thousand years ago, that hill was a beaver-shaped island swimming in glacial Lake Hitchcock. As the waters receded, the Wôbanakiak: Amiskwôlowôkoiak told a story of a giant beaver decapitated and left to form the low mountains of the ridge. My mother, the hydrogeologist, worked cleaning up the water ecosystems of Massachusetts. I looked at Beavertail Hill, phone in my hand, and I knew that the water of my body held toxic runoff, like the water of my home ocean, the Atlantic. A healthy future receded from me like a tide, and I reeled with vertigo. All at once, I fell into an oceanic whirlpool and watched myself being pulled down, as if a camera zoomed in on my face while the background fell away and blurred.

For the Fourth of July, just a week before the call from Dr. B, my boyfriend Ben took me to stay at his family's cabins on the Atlantic. His father and stepmother had taken us out on their boat, sailing far enough from the cove that the wind changed, the coastal smell of the shore replaced with a deeper, older scent. After the sun went down, our wake burned a luminous blue-green—bioluminescence, shining when the waves moved it, tingeing our throats, the undersides of our cheekbones. I'd seen bioluminescence before—the same blue-green. One night, at the cottage my grandmother rented on the Long Island Sound in the summers, my mother, my grandmother, my aunt, and I had gone to the water late at night. We'd peeled off our clothes, the four women of my family, left them on the sand, and run into the cold ocean in the dark, screaming with joy and the shock of the water. Around us as we swam, bioluminescence sparked the water, mirroring my delight in the night, the sea, my skin against the salt water.

That moment on the sailboat returned to me as I held the phone in my hand—the swirl of it, its fire and peace, confusing and healing and otherworldly. The bioluminescence took me both out of my body and into my body. I remembered that as the whirlpool drew me down.

"You should come in," Dr. B said over the phone. My body had disappeared, but I could feel my father behind me, his eyes on my back. "Come in on Monday, all right? We'll talk about it. I'm sorry. Tell your dad I'm sorry."

In dark places, it is light that calls us. Even in a broken ocean full of chemical fouling and warming water, bioluminescent fire remains. Sometimes, a warming and failing ocean makes it burn brighter. Spanish colonizers, naval officers tracking U-boats, and generations of sailors have seen this glow. We've written about bioluminescence for as long as we've kept natural history records, both as a phenomenon of mystery and as a support to human life.

Bioluminescence is a chemical reaction, usually produced by the combination of luciferin, a compound, and luciferase, an enzyme. It creates a glow within the body of an organism or outside of it in water when the chemicals are ejected. The bioluminescence I'd seen in the Atlantic was caused by dinoflagellates, a phytoplankton. They can appear in fresh water, but the majority, and the only ones that glow, are marine. Dinoflagellates photosynthesize, making energy from the sun like a good floating plant. But they also synthesize luciferin and create luciferase. Their luciferase is related to chlorophyll. Their bioluminescence, then, is a cold fire in water.

Dinoflagellates are essential to the marine ecosystem, a bedrock of the food chain, a marine alga eaten by equally tiny zooplankton, the floating animals, who are, in turn, eaten by larger creatures. Dinoflagellates who produce bioluminescence are often found in bays or estuaries, concentrated into narrow confines, creating a density in the water. In this density, they glow when disturbed. A 2017 study found that bioluminescence can reduce the grazing of copepods on dinoflagellates, making their glow a protection against predation.[1] Every stroke of a swimmer, dip of a canoe paddle, flash of a wave, wake of a boat, creates a nexus of light. Often, in the Atlantic, this fire is a pale blue, the light of stars, the Milky Way. Its poetry is in its visibility in the world.

And that light is sibling to other cold fires, so called because only 20 percent of the light emitted produces heat. Before the invention of safety lanterns, coal miners stashed fireflies in jars and took them underground, lights that would not explode in the deep pockets of the earth. In World War II, Japanese soldiers carried dried plankton, revived them with water, and in their glow read maps without drawing enemy fire. In the Indonesian jungle, travelers navigated by glowing fungus on the trees. To find your way where a simple candle might cause the air to explode, where the path will be too dark, where light could summon danger to your doorstep, you will follow any light. But bioluminescent light, created without fuel, reproducible, naturally available, is holy.

On the day Dr. B called and told me the news, Ben drove us to the top of the Holyoke Range, a line of east-west-marching basalt that defines the southern border of the inland New England valley. The ridge of the mountains was formed two hundred million years before, built by riffles of lava, and as this weathered, pillars of rock emerged, giving the hills a rusty color. He gave me a box of blueberries. We looked down on the Connecticut River winding to the Long Island Sound—my childhood estuary full of dinoflagellates—past tobacco fields, factories, nitrogen runoff. "I think it'll be OK," he said. "You've got time. Just look at your dad."

As he spoke, I stretched away from him. My father was on his second transplant, had survived longer than anyone else in his family. Ben hadn't been there for the first transplant, the deaths of my uncle and my aunt, the poverty, the terror, the chaos, my parents' brittle marriage.

Ben's mother died young. She said goodbye to him from the bed where she'd been since after the chemo stopped, just before he went out to the movies. She said, "I'm of a more Puritan soul," thinking he was breaking rules, going out instead of going to school, unaware it was a Saturday. When he came home, she was gone.

Her Puritan soul was on my mind. I would die young. All loyalty and goodness, he would stay with me. He'd go through all of it again.

"You remember when your mother came for her shoes?" Ben's father, Big Ben, asked him the night we sailed the ocean of stars.

"You mean when her ghost did?" Ben laughed.

"They washed right up next to us on the shore. We lost them after she died, you know? They washed off the boat," Big Ben said to me as he took a drink from the cap of a thermos filled with martinis. "And then there they were, on the shore." He looked at Ben, "Your mother came for her shoes."

I told myself that Ben's mother's death was why, from the moment I hung up the phone with Dr. B, I wanted to break up with him, like a hunger, down in my body, calculating and eerie.

I told myself I was protecting him. But I was protecting me.

I wasn't the girl he'd fallen in love with, who could have children and make a life with a summer house and a sailboat. I couldn't stand how he looked at me on the top of that ancient volcano, the same look my mother wore every day of my childhood, bound by love and duty to a fate that wasn't in her body but in my father's. She looked at him like a porcelain-cased bomb that could explode at any time. I'd seen her look at rivers the same way. She knew how sick healthy-looking things could be. Now she knew the same of me, and so did Ben. I wanted to get as far away from that look as I could.

Time has changed how humans use bioluminescence. In Vieques, Puerto Rico, there is a famous bioluminescent bay, Puerto Mosquito. Named for a pirate ship, and not for the insect, you can no longer swim in the bay. Swimming tourists and their sunscreen and DEET bug repellent have diminished its fire. At one point, the *Guinness Book of World Records* named it the most bioluminescent bay on Earth, with its native dinoflagellate, *Pyrodinium bahamense*, found in concentrations as high as 720,000 per gallon of water. Tour guides tell a story that the bay was corralled by the colonizing Spanish, who thought the light was the work of the devil. When they tried damming the glowing water to contain its danger, the increased concentration of plankton only made the bay glow brighter. When Hurricane Maria devastated the mangrove forests that protected the bay, the influx of water diluted the concentration of bioluminescent organisms and washed the dinoflagellates out to sea. For a time, there was little to no bioluminescence to be seen. As the island and the bay healed from the storm, the blue fire returned.

In 2018, a swirling, glowing whirlpool was detected by satellite in the Baltic Sea, so huge that it could contain Manhattan within its circumference. The influx of bioluminescent plankton multiplying in these heating waters is also turning it more hypoxic than they have been in 1,500 years. This bioluminescence indicates the illness of the sea, its movement toward collapse. In India that same summer, and for several summers after, overabundant algae glowed blue in the waters off Mumbai and Chennai, attracting tourists who came for the light, unaware that it indicated a dying ocean whose warmth encouraged deadly algal blooms. Scientists believe that warmer waters globally will make the oceans more bioluminescent as plankton respond with increased breeding.

In his office, Dr. B said: "I wish I could tell you some better news. That many cysts for someone of your age? It is pretty advanced."

"The family story is that it's harder on women."

"I don't know about that."

"My aunt was twenty-four when she went on dialysis."

"I think you'll have five years of good health before you lose kidney function, go on dialysis, go through menopause. You'll never have children."

"So. I'll be twenty-seven." I got quiet. "Lois died of a brain aneurysm at forty-nine."

"You've got an 80 percent higher chance of those than the general population."

"OK."

"This isn't what I wanted to be telling you."

My family drowned under the news. My father started drinking again. We went to dinner together. Sitting at the picnic table, the Holyoke Range in the distance, a roll of paper towels flapping in the wind, a glass tumbler of red wine in his hand, he said: "I'm not my PKD. You aren't either. I'm more than this thing."

"I don't know. I don't think I can separate it out."

"If you don't, it'll kill you."

My mother cried when she looked at me. She'd dress for work, spend the day testing groundwater, come home, hide in her bedroom reading sci-fi until she slept. Standing next to my parked car, summer darkness and the sounds of crickets surrounding the farmhouse, I'd see her light still on, and the Milky Way coursing over the house like swirling plankton.

I drove aimlessly up and down the coast, and the water let me know what to do. A few weeks before we left for college, I told Ben we had to break up and wept in bed afterward, sure I'd made a mistake from which I'd never recover.

Back in Seattle, I didn't call my parents, stopped speaking to friends, wrote a thesis on dying women, drove north to the Puget Sound and Deception Pass. I knew the Pacific was under enormous, generational, and existential threats, but I craved the euphoria of an animal surfacing from the gray-black water—a whale back, a seal's blocky head. I had no idea how to live with the grief for my body or for them, twinned and heavy. By the time I graduated the following summer, Ben was in love with someone new, and she was sweet and healthy and studying geology. In the picture Ben showed me, she looked like someone you could have a sailboat with.

I thought if I went back to where the ocean smelled right, where the waters glowed, I might find an answer. I bought a cheap car and packed it with books and a tent, and left the West Coast, singing sad songs with the windows open with Heather, the new friend I made in a seminar on the sublime. Heather, who couldn't drive, was the first friend I made in my new body, standing in the back of the Re-Bar, our favorite bar, seeing bands, transformed by music while I nursed a ginger ale, pretending I could make PKD go away by refusing sex, joy, alcohol, food. Her melancholy and separateness from people felt familiar. She handed me cassette tapes as I drove, smoked cigarettes with red lips and effortless cool, reminded me I was raised by grieving bohemians and had been admitted to the club of people who expect to die young.

Bioluminescence is no longer just a light in a jar. We use bioluminescence to test water quality and to indicate overheated oceans, ecosystems damaged by rising seas. We use it for health diagnostics and to research renewable energy sources. The first medical uses suggested for bioluminescence came from the Roman naturalist Pliny the Elder in his 77 CE book *Historia naturalis*. He wrote that the bioluminescent jellyfish we now call "night light of the sea," *Pelagia noctiluca*, could be boiled in wine and used to treat kidney stones and gout, both companions of PKD.

After the bloody Battle of Shiloh in the American Civil War, on a field littered with the bodies of injured men, there in the dark, some of their wounds became illuminated with a pale fire. The men whose wounds glowed recovered faster. Nurses and doctors at the scene called the phenomenon "angel's glow." In 2001, teen Civil War buffs working with scientists from the US Department of Agriculture learned that nematodes in the battlefield's soil had mixed with blood in the wounds, feeding bacteria until it glowed. That fire's antibiotic elements helped heal the soldiers. This same bioluminescence could be used as well to protect agricultural crops from the potato beetle.[2]

The Nobel Prize in Chemistry was given in 2008 to the discovery of a bioluminescent protein called green fluorescent protein (GFP) that can be found in the body of the crystal jellyfish, *Aequorea victoria*. Because this protein is also fluorescent, it becomes excited by blue light wavelengths and glows green. It can be inserted into cells to assist in diagnostics in disease and cell study. To assess heparin's efficacy, firefly luciferase has been used as an imaging agent in humans to track the presence of blood proteins, as well as in pregnant mice to investigate the nature of the blood-placenta barrier. Scientists have begun inserting bioluminescent genes into mammals to better understand illnesses like Parkinson's.

A naturally occurring bioluminescent bacteria, *Aliivibrio fischeri*, occurs symbiotically with Hawaiian bobtail squid. When submerged in water with high levels of toxicity, the respiratory functions of its bacteria are interrupted, causing their light to wink out, indicating a polluted waterway. It is hard not to ask the pale fire and the entities that create it to become our servants as we crave their magic, as they warn us, educate us, and protect us.

When I was diagnosed in 1993, I was told there was no cure, but every small spark is a clue, every shimmer in the black ocean a way to something more luminous and complex. In 1953, Jim Lovell—the savior of the Apollo 13 mission, who would visit the moon twice—was flying over the Pacific Ocean when his navigational gear failed. In the lowering twilight, he had to find the aircraft carrier, and he needed something to steer by. As he approached the carrier, he saw the glow of bioluminescent algae, stirred up by the boat's propellors, which guided him home.

When we graduated, Heather and I drove east from Seattle, pointing to the Atlantic. We stopped in national parks and a mission church in Idaho that ached with echoes of genocide. I was permeable to every broken thing, the vast wound of the West, hollowed out of animals and people and replaced with an empty simulacrum of place, or a place that was so obscured I couldn't read it at all.

We slept in small hotels. When Heather was in the bath, I'd lie on my bed and close my eyes, the whir of the road in my ears. I'd fall into a trance, like descriptions I'd read of deep meditation. Buffeted by fear, I wasn't in control. I lacked context for the mystery trying to find me, which felt bottomless, another whirlpool.

South Dakota rolled like waves, humming with the water it remembered, grass shifting as if in tidal pull and not in wind. We camped in the dry heat of the Badlands, and there was nothing to distract me from my feelings. The first night, we sat under the stars, both thinking about death in our own ways. Heather had lived with depression, suicidal ideation, and poverty all her life. We drank bottles of screw-top wine from a six-pack we bought at Wall Drug as trees swayed in the dry wash to our east, no campfire because of the drought.

"I don't ever want to leave here," she said, the first either of us had spoken in hours.

"I know. Me either." Without fire, we had nowhere to turn our focus. The dry land all around felt too large, vertiginous. But the pull of the fire of the stars overhead, shuddering behind wisps of clouds, drew our gaze as the flames might have. They wheeled over us, a time-lapse image. It was what I had been avoiding since diagnosis—the edge of the whirlpool.

"Sing something."

I sang Vic Chesnutt's "Dodge," his lyrics about leaving home echoing in the stillness: *I done shit everywhere that there is to eat. Guess it's time for me to get the fuck out of Dodge.*

I remembered Dr. B's office—*This isn't what I wanted to be telling you.*

In the silence after I sang, I thought I heard a bison snuffle. The darkness was near absolute, the sky overhead a rush of clouds. Heather took her eyes off it to look at me. "This is the first place I've been where I didn't want to kill myself."

"This is the first place I've been where I thought it might be OK if I died," I replied.

She handed me another bottle of wine. I cracked it and leaned back onto a blanket laid over the hard earth and watched the speeding clouds until I was too tired to watch any longer.

In the morning, the ground was alive with grasshoppers, a wave of them kicked up by every stride. They shushed out ahead of my feet, a bloom of insects born of drought, and I walked to a dry wash surrounded by brittle cottonwoods. They made the path an ocean, almost luminous, waves of grasshoppers on the dry prairie, like a flush of bioluminescent fireflies in the grass of a home meadow, the flash of dinoflagellates along the shorelines that raised me. On the horizon, a herd of bison grazed. That night, I went to sleep without watching the stars, exhausted by the holiness of the place. I dreamed that we were in a sailing ship on the ocean, hunting whales—I was a sailor, Heather was a sailor, we'd always known each other, over lifetimes.

A windstorm woke me up after midnight. The sides of the tent flattened over us in the gusts. I came to, unable to breathe, nylon covering my mouth. Heather slept on, and I unzipped the door and crawled sideways out into the cloudless South Dakota night, nearly knocked over by the wind. Randomly, in the full dark, I pulled at the sides of the tent, lacing an extra rope through the fly hooks on the top to keep it from blowing sideways, tying it to a windbreak picnic table we'd camped next to, pounding the tent pegs back into the packed earth. When the tent was steady, I looked up for the first time and saw the Milky Way stretching like a backbone over my head, listened alone in that wind to the sound of the distant trees, the whickering of horses from another camper's trailer. The universe was so much bigger than the fears I was carrying.

I thought: *What if the whirlpool I'm falling into is really a vortex that takes me out through the glowing oceans of the world and into the infinity of the stars? What if that's death, something bigger and wilder and more beautiful than the fear?*

I closed my eyes then, standing there, and felt exhausted, as if I were coming in from a long hike in the rain. All I wanted was sleep. The dead sea, the ocean of stars, the sea of grass, the memory of sailing—face illuminated blue with bioluminescence, ocean alight and alive—buzzed through me with something like an answer.

notes

1. Jenny Lindström, Wiebke Grebner, Kristie Rigby, and Erik Selander, "Effects of Predator Lipids on Dinoflagellate Defence Mechanisms—Increased Bioluminescence Capacity," *Nature: Scientific Reports*, October 12, 2017, https://www.ncbi.nlm.nih.gov/pmc/articles/PMC5638803/pdf/41598_2017_Article_13293.pdf.
2. Andrew Haynes, "Glowing Wounds and Angelic Bacteria," *Pharmaceutical Journal Online*, June 22, 2015, https://pharmaceutical-journal.com/article/opinion/glowing-wounds-and-angelic-bacteria; Sharon Durham, "Students May Have Answer for Faster-Healing Civil War Wounds That Glowed," *US Department of Agriculture Agricultural Research Online*, May 29, 2001, https://www.ars.usda.gov/news-events/news/research-news/2001/students-may-have-answer-for-faster-healing-civil-war-wounds-that-glowed/; James Byrne, "Divine Intervention via a Microbe," *Scientific American*, November 19, 2010, https://blogs.scientificamerican.com/guest-blog/divine-intervention-via-a-microbe/.

Fire and Psychoterratic Emotions

Glenn Albrecht

Fire is like water in that, in the spirit of Heraclitus, you cannot gaze into the same fire twice. Combustion is a process of constant change. Indeed, it is the protean nature of fire that fascinates us humans as we engage with the dancing flames, the glowing coals, and the curling smoke. Psychoterratic (psyche-Earth) emotions, both positive and negative, are what move or affect us, and the ancestral emotions of human contact with fire must have been largely positive. The human appreciation of fire has consistently been connected to comfort, food, entertainment, and security.

In the English language, though, there appear to be very few terms that describe positive emotional engagements with fire. I am sure that, like the Arctic Inuit people with names for different aspects of ice and snow, indigenous cultures all over the world have words for different aspects of fire. It would be useful to have a lexicon of such fire words and their meanings.

Perhaps, there is a need to put into the English language a name for a fire emotion that most humans feel but do not articulate. I suggest *reverfyre*, a state of particular reverie in the presence of a safe and convivial fire. A combination of the Old French *reverie* and Old English *fyr* (fire) is a start to rectifying the absence of fire-emotion concepts. I see reverfyre as a subset of *eutierria*, or the good Earth feeling of being at one with some aspect of the natural world.[1] It is to be hoped that good fire emotions can be returned as part of human culture in the Symbiocene, or period of symbiotic reintegration with the rest of life.[2]

The Indigenocene

There was a period in human history on planet Earth when humans were well integrated with the rest of life and the major elements—earth, air, fire and water. As a relatively newly evolved species, they really had no choice, as all these elements predated them by millions of years. It was a case of adaptation or extinction.

Before the Industrial Revolution and global-scale colonization by Europeans, including of the so-called New World, humans had spread from one or more nodal points of genesis (perhaps from many parts of Africa) to all parts of habitable Earth. As a mammalian species, *Homo sapiens* became a successful occupier of niches ranging from tropical forests to the Arctic Circle. Hence, as a mark of respect for endemic indigenous people, in situ the world over, a new period in human history needs to be named.

As a result of this evolutionary "success," I propose a new interval, the Indigenocene, starting within the geological epoch known as the Pleistocene and terminating within the Holocene epoch, which commenced 11,700 years ago and continues to the present. The Indigenocene, as a period of Earth and human history, commenced around 100,000 years ago (when humans had extended their range all over the planet) and began to disintegrate from the beginning of the sixteenth century AC (after Columbus). That point in history is important as it marks the so-called voyages of discovery by Europeans and the invasion of the rest of the world by the explorers, colonists, and evangelists that followed.

The term *Indigenocene* is derived from the Latin *indigena*, meaning "native" and "specific to a certain place or region." In the case of humans, *indigenous* means being a member of the original inhabitants of a particular region or territory. The suffix *-cene* is from the Greek *kainos*, or "new." The *Indigenocene*, then, is a term that implies successful, geographically distinct, adaptive diversity, not homogeneity.

The reason for marking this interval is that within the Pleistocene and Holocene epochs, humans—and no other land-based mammal species—successfully migrated (including by boat in long sea voyages) and adapted to all but one continent on Earth. However, in addition to adapting to what is now conventionally but erroneously called the "environment," *Homo sapiens* went on to engage in the active management of the flora and fauna endemic to the region they occupied. They were able to do this through the accumulation of knowledge, which was made possible by their intelligence, experience, language, technology, and cultural practices (drawing, dance, song, carving, sculpting, and so on). If there is a case for human exceptionalism in the class Mammalia, adaptability is surely the strongest one.

The Indigenocene reaches a point in time at which, by trial and error, and because of the accumulation of discovered knowledge, humans, the world over, can be said to be "native" to their home territories. They know their places' phenology, or patterns and rhythms; they are at one with the potential of the soil; they have developed rules and ethics to live within the community of life; and they have lore by which they can live relatively peacefully within the human community.

The net result of humans' integration into ecoregions was stability and a gradual increase in the human population, subject to limits imposed by the natural resource constraints of a locality and rules for cooperative living. Humans were symbiotically united to the rest of life, but they manifested that relationship in particular ways in diverse geographical locations. Without having detailed knowledge of contemporary symbiotic science, they understood their connectivity to other life forms in largely symbolic and animistic terms.

As Big Bill Neidjie (1913–2002), a member of the Buniti clan of the Gagudju people of the Arnhem Land region of the Northern Territory of Australia, put it in his wonderful 2002 book *Gagudju Man*:

> We walk on earth,
> We look after,
> like rainbow sitting on top.
> But something underneath,
> under the ground.
> We don't know.
> You don't know.[3]

One other crucial factor that enabled humans to become indigenous and endemic to particular places was the use of fire. Deliberate use of fire is one of the many features that make *Homo sapiens* unique in the animal world. The ability to create and use fire in locations of choice in order to manage a home symbioment and to cook a variety of foods is what enabled a recently evolved member of the ape family to become ubiquitous on Earth. Such is the human-fire bond that *Homo sapiens* could have been called *Homo pyrophilus*. Fire makes us human.

Fire Is Nothing

A human culture that has been extant for 80,000 years or more is found on the island continent of Australia. Before 1788, the start of British settlement, the accumulation of skill and knowledge in the use of fire had enabled up to a million Aboriginal people, who lived in hundreds of fire-prone bioregions, to live safely and harmoniously with fire.

As both Australia and Tasmania dried out and sea level rose during the late Pleistocene, the already-fire-prone, eucalyptus-dominated ecosystems became even more fire-prone. Australian Aboriginal people adapted to these changing circumstances and became *pyrophiles,* or fire lovers. They used fire to control wildfire. Without such intimate knowledge of all aspects of fire in the landscape, survival would have been difficult, if not impossible.

Today, Australian Indigenous people live within a bioregion susceptible to dry season fire, including in Arnhem Land in

Northern Australia. We can learn from them about the extent to which humans can take control of fire and make it a vital component of culture. Big Bill Neidjie, again from *Gagudju Man*, offered profound thoughts on fire:

> This earth I never damage.
> I look after.
> Fire is nothing, just clean up.
> When you burn, new grass coming up.
> That mean good animal soon.
> Might be goose, long-neck turtle, goanna, possum.
> Burn him off, new grass coming up, new life all over.[4]

How remarkable that someone could say in the early twenty-first century that "fire is nothing." So confident Big Bill was in the knowledge he and his culture had of fire and how to control it that fire was not only benign; it was an asset, vital to ongoing indigenous culture and endemic ecology.

Positive Fire Emotions

The use of "firestick farming," or the deliberate use of fire for ambush hunting and to maximize the productivity and utility of the land for humans prior to colonization, meant that in order for fire not to disturb the peace, communication of one's intentions was critical. The awareness of the benefits and burdens of setting fire to "country" was a primary ethical consideration to ensure the well-being of others. However, fire had many other functions in traditional society, as well:

- Having fun!
- Providing warmth
- Using for cooking
- Providing security (from predators and enemies)
- Signaling and communicating
- Clearing the land (create "roads" through thick bush)

- Killing snakes (many poisonous and edible varieties)
- Hunting (ambush with fire)
- Supporting regeneration of plant foods
- Extending human habitat zones (e.g., burning rainforest in Tasmania)
- Gathering information (what is burning and where?)

The use of fire as an integrating force in human social affairs is illustrated by W. E. H. Stanner in his 1976 essay "Aborigines and Australian Society." Stanner imagines a story of interrelated harmony between fire and humans as he explains the maintenance of social order in traditionally configured tribes. The concept of a fire pattern is tied to human order: "The band was in camp for the night. The camp took the shape of a rough circle perhaps twenty paces in diameter. The circle was marked out by a ring of fires. The fires and the sexes alternated in an interesting way: if there was a man on one side of a fire, there was a man on the other side; if there was a woman on one side, there was a woman on the other. The fire-pattern of the camp thus revealed an underlying social pattern."[5]

Big Bill's statement "fire is nothing," then, did not mean that fire had no impact on human affairs: in both nature and culture, humans control fire to establish and maintain forms of order. Fire became part of culture, with its control and management integral to establishing and maintaining relationships within clans and those of adjoining regions. John Bradley, in his 1995 case study on the emotional importance of fire in the Gulf of Carpentaria and its offshore islands in Arnhem Land, highlighted the powerful emotional impact (that is, being moved) of fire. He relates the testimony of a female elder, Ida Ninganga, who "spoke in raptured and passionate tones" about smoke mail observed rising from the islands: "Oh, all of the islands, they would once be burning, from north, south and east and west, they would be burning, the smoke would be rising upwards for days, oh it was good, you could see the smoke rising from here and also from Borroloola, you knew where all the families were, it was really good, in the times when the old people were alive."[6]

I created the concept of "endemophilia" (the particular love of people for the endemic distinctiveness of their country) as a positive psychoterratic human emotion similar to that experienced by Ida. For example, the experience of smoke rising was a sign that not only were people well; they were actively improving their habitat. The coincidence of ecosystem health and human physical and mental health was a sign of total health. However, as we will see, the emotional polarity of fire can be reversed.

The Arrival of the Anthropocene

With the systematic colonization of indigenous land and peoples by European nation-states from the sixteenth century onward, the desolation of Indigenous cultures commenced and the intentional and unintentional acclimatization of "foreign" ideas and species began to overwhelm the endemic. Slavery, invasion, genocide, colonialism, imperialism, "development," extinction, and imposition of the multiple despotic ways of a dominant Eurocentric power structure proceeded apace. The physical, cultural, and psychic toll on people was profound and reverberates into the present.

As the Eurocentric form of economic development (capitalism) became global in scale, so did its negative impacts. So powerful were the humans who controlled the levers of the industrial form of economic growth that it seemed humans had exited the Holocene.

The idea of a new geological epoch, the Anthropocene (as yet to be ratified), was originally forwarded by Paul Crutzen and Eugene Stoermer.[7] They argued that a new epoch, one characterized by humans—*anthropos*—dominating all relevant biogeochemical cycles on Earth, needed to be added to Earth history. The stability of the Holocene was gone, and a new era of climate chaos and biological extinction commenced.

It is arguable that the Anthropocene proper commenced in 1950 and that, by then, the Indigenocene as a coherent "age" had been systematically extinguished, as hardly any native culture remained "untouched" by so-called Western civilization. In 1950 there came the great acceleration of economic growth and the Western model of development. The Anthropocene, with its homogenizing gigantism, had erased the Indigenocene and its heterogeneous diversity by the end of the twentieth century.

The acceleration of capitalism, this extractive and polluting form of economy, caused changes to the biophysical parameters of life on Earth. Global climate warming, the addition of radionuclides, ubiquitous plastic pollution, biological extinction, and the overshooting of many other vital biogeochemical systems for life on Earth were delivering an unsafe operating space (lifeworld) for all life, including humans.

Solastalgic Fire

The forced cessation of human control over fire had a significant impact on Aboriginal Australians, creating a new unsafe cultural operating space. On the impact of colonial dislocation on the loss of cultural fire practices, I defer to Dean Munuggullumurr Yibarbuk of the Maningrida township in Arnhem Land. In 1989, he stated: "Today fire is not being well looked after. Some people, especially younger people who don't know better or who don't care, sometimes just chuck matches anywhere without thinking of the law and culture of respect that we have for fire."[8]

The change from nonpermanent settlements in the area of Arnhem Land to larger permanent settlements (such as Maningrida) meant that, for people in Indigenous towns, relationships to all places became tenuous. Those still living semi-traditional lives at Outstations were able to continue controlling wildfire via what has been called *cultural burning,* which involves burning only in cool seasons and when it is safe to do so (such as in conditions of green vegetation, low wind, high humidity, and low human risk). However, sedentary people in towns generally lost close contact with both the knowledge and the praxis of managing fire. Elsewhere in Australia, the forced relocation of Aboriginal people to towns or their "pressured" migration to cities meant that cultural burning was in danger of becoming a lost science and art. If control of fire in Arnhem Land was in trouble, then it was in trouble across the whole continent.

The collapse of traditional life for Indigenous Australians was and continues to be both geographical and cultural. Loss of place and the lack of order in "new" places presaged my concept of "solastalgia." I have described the emotion or psychological state of solastalgia as the lived experience of negative environmental change, mainly in relation to distress associated with the desolation of one's home environment by the imposition of activities such as mining. Unwelcome change to a "place" by forces so powerful that individuals and whole communities cannot stop them was foundational to my definition of the emotion of solastalgia.

Although I have defined solastalgia as caused by a chronic change agent, such as mining or climate change, the effects of wildfire also have elements of this type of change. After the immediate impact of large-scale wildfire, even if people had been able to evacuate safely from their home environment, they return to a newly devastated state (even if their houses remained intact) and have to live with a profoundly changed landscape for years. The world over, there are now political and psychoterratic responses to the solastalgia induced by wildfire.[9]

Fire Is Everything

In what has been referred to as the Black Summer of 2019–2020 in Eastern Australia, the emotions of fire became negative for most Australians. Any sense of human control over fire was razed in conditions described as "catastrophic" by the authorities, who give warnings about what humans can expect when drought, strong winds, and heat coincide with wildfire.

Experts have estimated that millions of vertebrate animals were burnt in the Black Summer bushfires in Australia. Trillions of invertebrates must have been immolated as well. In the Australian context, this meant the remains of plants and creatures great and small were constitutive of the palls of smoke and, in combination with wildfire, created pyrocumulonimbus clouds, complete with their own ferocious lightning and storms.

Wildfire was becoming the "new abnormal," immolating human and nonhuman life. So abnormal were conditions that alpine places, such as the Central Plateau of Tasmania, never before burned in known geological history, were lost to wildfire. In addition, wildfire was occurring in the winter season, something new to settler-colonial society.

Climate calescence (warming) has also entered the picture as a new form of colonialism. Australia is one of the world's leading exporters of black coal. Unfortunately, as a result, it imports anthropogenic climate change to an already-vulnerable landscape. Australia is not only a willing contributor to the Anthropocene; it is within its late stages, in an era the fire ecologist Stephen J. Pyne has named the Pyrocene (age of fire).

As the planet heats and the risk of catastrophic fire ramps up, there is hardly a place on Earth not at risk from wildfire. As we edge ever closer to tipping points in the climate, never before experienced emotions are rising from the Promethean depths of our individual and collective psyches. No longer is fire connected to the creation of order; it is now firmly tied to the creation of disorder. We all now live in unsafe cultural operating spaces that are having profound impacts on our psychic and physical health.

The Negative Emotions of Fire

As I warned earlier, the emotional polarity of fire has been reversed, and all humans are likely to experience new negative fire emotions. The wisdom of Indigenous elders is no longer as tightly held as it was in the Indigenocene, and even where it remains, local knowledge is being made redundant by the rapid movement of ecosystems further north and south of the Equator as the planet heats. The relevant "location" of local knowledge has migrated and uncontrolled fire is spreading in the age of solastalgia.

In addition to solastalgia, we now must engage with a range of negative psychoterratic emotions (from mild to extreme) to understand our predicament in the Pyrocene.

Ecoanxiety

Where solastalgia is the lived experience of actual negative change to a loved home environment, ecoanxiety is a form of anxiety connected to a person's uncertainty about the future state of an ecosystem and the possibility that it will become hostile to their needs. It is a state of uneasiness, dread, and apprehension about an uncertain future. The ongoing failures of international forums such as the UN Conference of Parties to deliver tangible reductions in greenhouse gases only add to society's cumulative anxiety. With each annual incremental increase, collective anxiety about the climate heightens. Climate scientists, in particular, feel this form of anxiety more than most.

Meteoranxiety is a subset of ecoanxiety that refers to the particular kind of anxiety one gets from the extremes that are now being experienced in weather patterns. As climate calescence increases, regular new extremes (droughts) and records (temperature and rainfall) in the weather mean that it is no longer possible to be able to anticipate the impact of weather events by referencing past norms.

Anxiety about the next extreme weather event increases with the use of information technologies such as satellites, which generate data and help craft forecasts delivered to 24/7 weather channels and personal mobile phones. In many instances, anxiety about the possibility of being hit by an extreme event such as a superstorm is itself intensified when one can "see" it on a smartphone screen heading straight toward a particular location.

Pyroanxiety

In addition to smoke on the horizon and in the nostrils, pyroanxiety is now delivered by instant warnings and maps on computers and smartphones. I find that the terms *ecoanxiety* and *climate anxiety* are too imprecise to accurately describe the type of anxiety that builds on hot, windy days in fire-prone areas. I have felt this type of anxiety on a 47 degree Celsius day at Wallaby Farm in New South Wales as I got the firefighting equipment ready for a possible battle against an evil "ember attack."

Pyrotrauma

Finally, wildfire creates pyrotrauma, that moment when you experience the sudden, hot breath of fire invading your personal space. Pyrotrauma is a subset of *tierratrauma*, a condition in which an acute traumatic event takes place to the psyche as a result of a sudden change to one's immediate surroundings. Being burned to death is an unimaginably terrifying event, and for survivors, severe burns to flesh are torturous. Fire scars are not only the damage done to tree trunks as fire passes through a forest but also the psychoterratic scars on humans who seek psychic and cultural healing responses to fire trauma.

Fire in the Symbiocene

I have argued for an urgent exit from the Anthropocene and rapid entry into the Symbiocene. Actively creating the circumstances in which humans once again engage in the project of life with other life forms will require novel forms of fire culture. While elements of Indigenocene fire culture remain hugely relevant, the new catastrophic climatic conditions in the Pyrocene will force the use of technologically assisted transnational and transborder collaboration on firefighting (with equipment such as water bombers) until such time as bioregional stability and culturally informed fire management merge. At that point, nature and culture will once again exist within safe operating spaces.

At this good turning point, emergent hybrid fire control will integrate traditional ecological knowledge and new scientifically tested methods to turn fire back into nothing—"just clean up," as Big Bill Neidjie said—and to reintegrate control of fire into new cultural norms.[10]

Ironically, control of fire will be returned to globalized humanity when we stop burning and exploding the fossil fuels of the Anthropocene. Only then, as we set fire to that renewable resource, recently dead wood, will we rediscover that primordial positive emotional state of reverfyre, and in companionship with other humans in front of a roaring fire.

notes

1. *Eutierria* and other psychoterratic terms I use to highlight the human emotional engagement with the state of Earth (good or bad) can be found in my book *Earth Emotions* and in the glossary available at my blog *Psychoterratica*. See "Glossary of New Words for a New World," *Psychoterratica* (blog), https://glennaalbrecht.wordpress.com/2017/12/23/new-words-for-a-new-world/.
2. Glenn Albrecht, "Exiting the Anthropocene and Entering the Symbiocene," *Minding Nature* 9, no. 2 (2016): 12–16, https://www.humansandnature.org/filebin/pdf/minding_nature/may_2016/Albrecht_May2016.pdf.
3. Bill Neidjie, *Gagudju Man: Bill Neidjie* (Marleston, Australia: JB Books, 2002), 75.
4. Neidjie, 18.
5. W. E. H. Stanner, *The Dreaming and Other Essays* (Melbourne, Australia: Black Inc., 2009), 251.
6. John Bradley, "Fire: Emotion and Politics, a Yanyuwa Case Study," in *Country in Flames: Proceedings of the 1994 Symposium on Biodiversity and Fire in North Australia*, ed. D. B. Rose (Canberra: Biodiversity Unit, Department of the Environment, 1995), 26.
7. Paul J. Crutzen and Eugene F. Stoermer, "The Anthropocene," *International Geosphere-Biosphere Program Newsletter* 41 (2000): 17–18.
8. Dean Yibarbuk, "Introductory Essay," in Marcia Langton, *Burning Questions: Emerging Environmental Issues for Indigenous Peoples in Northern Australia* (Darwin, Australia: Centre for Indigenous Natural and Cultural Resource Management, Northern Territory University, 1998), 5.
9. Erica Tom, Melinda M. Adams, and Ron W. Goode, "Solastalgia to Soliphilia: Cultural Fire, Climate Change, and Indigenous Healing," *Ecopsychology*, May 15, 2023, http://doi.org/10.1089/eco.2022.0085.
10. Tom, Adams, and Goode.

Simile

David Baker

1.

Orange-and-midnight the moth on the fringe tree—
first it nags a bloom; sips and chews; then shakes
the big flower. Then its wings slow. Grows
satiate, as in sex. Then still, as the good sleep after.
Each bloom a white torch more than a tree's flower.
Each is one of ten or twelve, conic, one of many
made of many green-white or white petals
held out, as by a hand, from the reach of the limb.
A field this morning was full of white moths. More
in the side yard, in the bluebottle, lifting—fog
off the dew, white wings like paper over flames
and floating awry or pieces of petal torn off.
Weeks now my words on paper have burned.
Burned and flown, like a soul on fire, with
nothing to show but ash, and the ash flies, too.

2.

Today, in the news—so many martyrs—
an "unnamed suicide bomber" took herself into
the arms of flame, and five others, "by her own hand."
Whitman means the beauty of the mind is terror.
Do you think I could walk pleasantly and
well-suited toward annihilation?
But there is no likeness beyond her body
in flames, for its moment, no matter its moment.
Yet the fringe bloom burns. Yet the moth shakes
and chews, as in sex. When the young maple
grows covered with seeds, they are a thousand
green wings, like chain upon chain of keys,
each with its tiny spark trying the black lock.
A tumbler turns and clicks. The world once more
fills with fire, and the body, like ash, is ash.

Fire in the Belly

Tyson Yunkaporta

The potential of a gut-brain, fire-lightning connection is a constant process of inquiry for me, a hypothesis that is neither verifiable nor falsifiable but that still demands my intellectual engagement because of the peculiarities of my totemic relations on the continent currently known as Australia. My fatherside totems (Apalech clan, Western Cape York) relate to a monsoon season of storms, and my motherside totems relate to a cold season of abundance and bushfire. I take my fatherside ways, but no man works with fire on the land without respecting the authority of his motherside family, as with most things. Even in patrilineal cultures, matriarchy is still strong, and fire sticks work only if one is male and one is female. So motherside totems, such as sparrow hawk, are at the front of my mind when I tell stories about fire.

Sparrow hawks spread bushfire by snatching burning sticks from our campfires and dropping them in the dry grass, blown flat by seasonal winds, to scare up small game. This is part of the pattern followed while caring for the land with fire-stick management practices performed by humans as a custodial species in the region. Fire is connective across domains in spirit and practice, through marriage and ritual. The regenerative friction between male and female energies and the dynamic places and people in which they overlap keep fire burning in the universe.

In my family, Brolga is a crane associated with a pantheon of animal and plant species, substances, and phenomena: blood, urine, Rainbow Serpent, feet, whirlwind, lightning. Brolga and lightning share the same totemic classification in our science, but fire and lightning are only indirectly related. Friction between the earth and sky creates a charge of connective power in lightning bolts that can ignite vegetation and turn sand to glass. Fire and lightning are only tangentially connected in this cosmology, an indirect relation moderated in my case by marriage across ritual complexes in the same way that the gut and brain are moderated by the strange functions of the cranial nerve complex.

The way I learned to set fires on Country was always through the regenerative disruption of everyday life processes that humans conduct when seeking sustenance and resources on the land. To find honey, an area might be burned to make walking easier in thickly grassed savanna at precisely the time that soil needs the ash and stimulating flame to begin its next regenerative cycle. Just as when collecting fibers for string or fishing at a precise time of peak fat in a coastal bioregion may require the burning of grass to remove the annoyance of mosquitoes, the grass seeds there need to be activated by the fire's heat and smoke. Symbioses have developed over deep time, with affordances built into the landscape as indirect communication between land and people that indicate what is needed by an ecosystem in which humans are still embedded within appropriate ecological niches.

Being separated from Country in the city for several years now, this communion is no longer apparent to me in relation to my habitat. The focus of my inquiry into the nebulous connection between earthly flame and celestial static has been driven inward. Land and kin relations no longer moderate my praxis of sparking and grounding cognition, which has become a hazy and stochastic process that is far too conscious to be effective. I am overthinking and underfeeling my existence, making inaccurate predictions from incomplete data sets with increasingly dire consequences. Being too much in my head has given me piles (hemorrhoids) and irritable bowel syndrome as my gut stagnates and atrophies in isolation from the rest of my existence.

Executive function is a powerful neural process that should never be deployed by individuals acting outside of community and the Law of the land. In Australia, we learned this the hard way millennia ago. Much of our Lore recalls stories of people transgressing in this way and their subsequent crimes, punishments, and learning. The adolescent cultures of modernity and postmodernity are beginning to relearn such lessons, long forgotten in industrialized lineages. The myriad sparks of individual sovereignty tend to incinerate a place and people after a few centuries, and the slow burn of collective self-determination is a regenerative pattern that keeps creation in motion indefinitely.

Frontal lobes are helpful when processing the cognitive dissonance that arises when sharing agency and resources with diverse beings existing in interdependent relation, but when they are deployed for audacious bootstrapping in an environment of hypercompetitive behavior, the results can be catastrophic. Individualistic thinking is never enough to sustain life. That's why our custodial species has developed two brains—one for thought and one for thinking-feeling.

There is a "big power" in your gut that must be attended to vigorously, in connection with the bellies of all others with whom you are in relation. It not only burns your food to power every cell in your body; it exists independently of your central nervous system, unlike every other part of your body. The neural-like processes of this sacred system process dark data and incomplete data sets using not cognition but logics that resemble what we call emotion today. The separation of thought and feeling in recent centuries has had a dramatic impact on the functioning of the belly. Rigorous gut feeling is supposed to have veto power over your executive function. If ignored for long periods, it will seek ways to sabotage your murderous path of narcissism and bring you back into proper relation with the beings and lands that sustain you.

The human two-minded system of brain and belly mirrors the cosmological structure of existence on this planet. In the turnaround event of creation, sky-camp and earth-camp separated, forming a complimentary dyad—as above, so below. Earth and sky, powered by the fires of their interaction, form a sentient system. Stars move in patterns that signal activation of seasons and activity for every terrestrial agent, including sentient bioregions. There is Lore and knowledge up there, as there is down here. Terrestrial agency, like the operation of the gut, is more fluid than the dance of stars so that systems may be responsive in real time to the inevitable transitions and upheavals of creation.

This agility is possible only when every agent in an Earth-based system is operating independently while responding to interdependent relations. The horrendous complexity of this responsive ontology defies mechanical computation and so must be thought-felt through many sentient bellies networked together in distributed, autonomous collectives. The governance of this collective executive function must be grounded in the Lore (story) and Law (protocol) of the land. The Law is in the code, the essence of sacred entities that exist as potentialities in the earth and emerge during dramatic creation events. They are maintained in sacred sites by humans as custodial beings, and those sites produce flows of energy that keep creation in motion.

The flows from sacred places in the earth and sky foster negative entropy, offsetting the inevitable breakdown of complexity and loss of energy that occurs in all systems as a result of what many call the second law of thermodynamics. A more human way to express this concept might be "It's easier to break shit than make shit." All things expire over time under this law of physics, which is what gives rise to the concept of linear time in cosmologies of singular entities and closed systems, cosmologies that no longer care for the energetic flows of sacred sites that connect land and sky.

More cyclic models of time suggest that entropy is not bound by beginnings and endings in regenerative systems. Energetic feedback loops between the earth and sky and all the systems within these domains maintain the infinite nature of creation. The fixing of nitrogen in soil by lightning strikes is a measurable example of this process. The death and the waste of one system is always another system's lunch, which is why shit is sacred in the Lore of the land. Your brain and gut relation mirrors those cosmological processes.

The gut governs terrestrial relations and is in constant communication with land and all our human and nonhuman kin. The head follows slower signals and cannot pivot in ways that are informed by complexity, so the mind can manifest as a top-down governance relation when it is not used in concert with the gut. When a person's head is in the clouds, carrying or longing for unearned power and privilege, their gut is out of right relation, which may result in a series of cascading failures that affect all the systems and beings around them. Executive function should always be signed off by the gut before any action is taken, if these consequences are to be avoided. Your gut might pick up the subtle informatics of a car salesman weirdly staring at your kids while you're distracted and checking under the hood—so your head may be telling you it's a good deal but your gut is telling you not to trust this creep.

The metabolic flames of the gut can burn hot, cold, or just right depending on how the fire is tended in relation to constantly shifting contexts in land-based and social dynamics. It must be kept clear in order to function correctly. Unfortunately, this involves behaviors widely regarded as dysfunctional or pathological in contemporary contexts. Outbursts allowing full expression of rage, annoyance, fear, joy, and grief must be allowed to run their course in order to clear the big spirit that rests in the belly and maintain it in a constant state of flow, in which energy enters, swirls, gathers, and is dispelled in regenerative patterns of behavior.

I don't do this anymore. I am at risk of ostracization and imprisonment if I yell "M———!" in the street or throw things around in my tiny new living space. (I can't put that word in print here as it is highly inappropriate, but it's the one that rhymes with *brotherfucker*.) As a self-managing neoliberal subject in a growth-based economic system, I am responsible for managing my behavior and a tidy workspace, minimizing conflict and violence, which includes trying not to use profanity in the content I produce for settler audiences.

I'm not very good at this job, a fact that renders my continued existence precarious. The subsistence of my family depends on my ongoing ability to engage in the production of text and audio that stimulates thought in ways that keep the wheels of capital turning. Without daily signals from land and sky, the machinery producing these narratives is contained only within my mind-body system, which I'm currently driving like I stole it to cover rent and living expenses.

Systems in isolation (like rogue predators, dying suns, settler campfires, and transphobic celebrities) can enjoy short periods of hot-burning illumination but ultimately collapse into ashes, cold, and darkness. Without the complex interconnection of a thousand regenerative loops, every engine breaks down after a brief, flaring L-curve of production. Artificial systems bent to singular goals (like empires, multilevel marketing schemes, celebrity sex tapes, and solar pool heaters) can never harness the economies of scale that are present in nature, the metabolic efficiencies of "energy return on energy invested" in balanced ecologies, where the aims are manifold and the outcomes distributed across and between systems.

I'm getting horribly fat. This is not an aesthetic judgment but a measure of my ability to continue working and keeping my family alive. Tying shoelaces makes me pant like a dog now, and I'm so exhausted most days that I'm experiencing sleep apnea while awake, as I regularly forget to breathe. Respiration has become a conscious process for me, but this is part of the cognitive load that comes with obesity. Fat is beautiful, sure, but it feels like shit. As an isolated organism, I am not harnessing the regenerative efficiencies of scale that occur in healthy habitats. I reflect the health of the systems I inhabit, producing less and less energy in relation to the fuel I burn, stockpiling toxic reserves until the wheels stop turning completely.

Usually, increasing the size of an organism will also increase the efficiency of its metabolism, raising its longevity and outputs. In the tinkered systems of modernity, these economies of scale are greatly reduced. At my waistline's current rate of expansion, I can expect to double in size by next October, but the fire in my belly that powers my existence will not burn 25 percent brighter. Growth is good if it comes from pregnancy, but not so good if it comes from doughnuts.

I'm not suggesting that this is true for all people of girth—perhaps there are many who enjoy more interdependent relations with their environment and therefore achieve the same economies of scale present in balanced ecologies, such as Israel Kamakawiwoʻole, that Hawaiian fella who played "Over the Rainbow" on the ukelele. He looked about fifty times healthier than Meatloaf, although respiratory failure did take him in the end. There are always limits to scale, as everybody from blue whales to Jeff Bezos knows.

A rat enjoys a longer life span than a mouse half its size because its metabolic fires burn at a rate of buy four, get one free, reducing its heart rate. Both lives have the same number of heartbeats, but the machine-gun pace of a mouse's cardiology makes for a shorter life span. I can feel my heartbeat in my face like a death-metal baseline right now, and I know I'm the caged mouse rather than the free-range rat in this scenario.

The quality of fuel is almost as important as the efficiency of the mechanism. Every habitat offers the perfect kind of wood in perfect quantities for small campfires, wood that cannot be used until it has fallen from the tree or bush. If you are present in the right season to harvest the appropriate resources, the optimal fuel will be present. There are affordances in Country that tell you where to be and what to do, and cooking or keeping warm is always provided for in right relation with the land. Your fuel is dry, your fuel is dense, and it is always in a location convenient to your fire, if you have placed your camp in response to the needs of the land. The land must also eat and metabolize your waste—your food scraps, shells, bone, flesh, urine, and excrement. Of course, this should never occur in proximity to the source of your drinking water, which is why the firewood is always useless near a creek or pond.

In times of disruption or emergency, you may have to burn unseasoned or wet wood. There are ways to build that fire to minimize the amount of toxic smoke in the air and produce enough heat to dry and warm your children and cook their food. However, in our culture, this is supposed to be a rare occurrence in response to urgent need. You don't burn green wood if you can avoid it.

We have a firepit in the small yard at the front of the unit we are renting. Recently, we bought a bag of wood from a service station and that wood was green. It was also the wrong wood for a fire in this place and season on Boonwurrung land. We built the fire carefully to dry out the sap that bubbled from the ends of each piece, but still there was dirty gray and black smoke in the air. The wood burned briefly and cracked into ugly charcoal pieces that immediately went cold. It would hold heat only if new wood was constantly burning on top of it, very much like the economic and social systems that have been imposed on this place. Our campfire, like so many contemporary economies, had a slow metabolism and voracious need to consume fuel and stockpile useless charcoal. It is out of place and out of time.

My thinking and my narratives are similarly complicated and resource-intensive these days. This is a result of writing rather than speaking. Literacy rewires the brain in catastrophic ways at the biological level, thickening the connective pathways between the logical and intuitive hemispheres of that organ and forcing some neural functions to migrate between one side and the other, squatting in territories where they don't belong. Our facial recognition mechanism, for example, is paying rent in a hostile neighbourhood where it can barely function. These facts I can back up with scientific evidence, however I can only speculate on the impact that print has on my gut-brain connection and my ability to think-feel my way through this world.

If your story is wrong, your thinking is wrong. Disconnected relations with place and entities both human and nonhuman can result only in crazy thinking, from my cultural perspective. However, that is knowledge that defies measurement and can't be taken on faith anymore. There is a lot of crazy thinking in the world, spreading faster than wildfires and devastating our living systems worse than superstorms. Crazy thinking is often presented as ancient wisdom, because the content is needed as more fuel for the fires of pseudointellectual consumption. Ayahuasca retreats. Paleo diets. Indigenous mindfulness workshops. Anti-vax elders on YouTube. Native corporate-coaching consultants. That auntie (and we've all got one) who likes to whisper into microphones, "There's no justice—there's just us." The Ancient Wisdom Industry pays well, so grifters abound.

My spouse and I work toward a time 453 days from now, when we hope to have the resources and freedom to move to her Country in central Queensland so our children will grow up to know their heritage and the right way to live. We will be closer to my family's Country there, a little farther to the north, so they will be able to understand their fatherside ways as well. I hold hope that they'll be able to reconcile their brain and belly, north and south, lightning and fire, father and mother thinking. However, we struggle to find a location on her land that is not ravaged by mining, ruptured by the extraction of coal and gas to fuel the fires of industry and technology. My own family's land is becoming similarly disrupted by bauxite extraction for production of aluminium, which will probably be used as an alternative for electric wire when the last of the world's copper reserves are gone.

Energy infrastructure lasts only thirty years or so, and power lines must be replaced regularly no matter how many windmills and solar panels you install, a process requiring trillions of tons of metal. Either way, the motherside and fatherside lands of my children will both be consumed by the fire that produces power and the flimsy metal required to conduct it.

I have greater concerns about phosphorus, however, most of which has washed into the sea after fueling the "green revolution" that allowed the intensification of agriculture needed to keep pace with the exponential growth of industry, and the human labour that needed to be burnt to power it. There are no new reserves of phosphorus to be found now, and it is difficult to grow plants without it. It does you no good to have a clear, strong metabolic fire in the belly if there's nothing left in the world to eat.

Still, many dream of a glorious future fueled by human sacrifice and ingenuity, an age of rockets and robot slaves driven by the flames of imagination and the power of positive thinking. I find it hard to share that dream, but that might be a result of my obesity-driven depression. As a human, I am hardwired to respond to my habitat and reflect its condition in my mind and body, and in that respect, I am a healthy example of my species. I should relax into my condition and stop overthinking. Too much lightning, not enough fire. Too much brain. Cognition is exhausting under these circumstances, so I outsource half of my neural function to a tinkered electrical system of automation to process my words, finish my story for me.

When I close this computation mechanism that I've outsourced my knowledge to, then reconnect with the green wood fire belching black smoke in my belly, I will see only a wrong story drifting up into a cloud that does not move or dissipate as clouds should. I wonder what will occur when those rains come, although I no longer have the capacity to predict such outcomes with any accuracy. If my gut were working, it might tell me to move to higher ground while I still can.

But there are embers in there still, if I sit quietly and rub my hands together then rest them just beneath my navel. My gut whispers, "How many Indigenous thinkers does it take to change a light bulb?" and I guffaw until it hurts. It's hard to overthink things when you've just laughed your head off. That must be why they call it a belly laugh.

I Vow to Be the Small Flame

Tamiko Beyer

My people, we have found
too shallow our roots
in this land full of boulders.

But when the satellites fall
I vow to use my good
sense of direction to find you.

Songs make provisions—
all the spells to turn our capillaries
into branches—

sea waving sky. I vow
a ravenous undoing.
I vow to love the fire always.

Mentorship from Fire

José González

When I first envisioned the work I wanted to do by creating what became the community and organization Latino Outdoors, and later through equity and inclusion facilitation in the conservation field, I did not anticipate being mentored by fire. In fact, I would say I was fairly fire-averse. Yet as I have become receptive to mentorship from the land and incorporate a life logic with the language of ecology and fire, along with other elements and nonhuman kin, the land has guided my work and personal cultural wayfinding. Land's mentorship has directly informed how I think about our sociocultural landscape. Through that, fire became a mentor to me as well, offering insightful answers to questions such as, What have we inherited that no longer serves us? and What must be burned to nurture a healthy landscape?

Things did not start this way. As a kid in Mexico, fire served a very utilitarian role of cooking food in my abuela's kitchen or out in the milpas of the rancho. I do not recall thinking too deeply about it. Fire was a tool—to be respected, sure, but to be used.

Years later, after migrating to the United States and growing up in the suburbs, I recognized that my family did not use fire in quite the same way, although its central role of utility remained.

Still, my exploration of fire expanded.

In high school biology class, we tangled with the question, What is alive? Part of this question was based in the categorizing approach of Western science and thus core to the purpose of the class, but at the same time, the philosophical aspect of the question intrigued me. We would ask, Is fire alive? It reproduces, consumes, moves, and so on, meeting most, if not quite all, the "conditions" of life. I think no one was seriously making a case for fire being alive, but it sure was fun to play with the idea and use it as an example. Little did I know that a few years later, I would revisit this question not just from a biological perspective but from a cultural and spiritual one as well.

As an undergrad, I yearned to connect with the different aspects of who I was. In a practice of deep cultural wayfinding, I reconnected to Indigeneity through my connection to Chicanismo—and with that came a reintroduction to the role of fire in ceremony. For example, fire played a role as a catalyst for different types of *limpias* and aspects of *curanderismo,* and it was a clear foundational element of sweat lodges. As Erika Buenaflor notes in *Cleansing Rites of Curanderismo: Limpias Espirituales of Ancient Mesoamerican Shamans,* fire "is one of the most common limpia tools" and can be used to "create a path toward a more graceful and positive transformation and to mark a new beginning."[1]

For those unfamiliar with *limpias,* the word comes from the Spanish for "clean" or "to cleanse." In a related way to how we might approach a digestive cleanse with juices and the like, a *limpia* is about a spiritual cleanse, removing bad energies and blockages of good energy. *Curanderismo* is the practice that includes these cleanses, especially those led by *curanderos* and *curanderas* who still retain ancestral practices. For example, one aspect of *curanderismo* that has entered the popular mainstream is "smudging" with sage to cast bad energy out from a place. Yet like all traditional and ancestral knowledge, it is important to anchor such practices to their cultural roots and the relationship of respect and reciprocity with the land.

I recall the first time I was invited to a *temazcal,* a sweat lodge and ceremony grounded in Nahuatl tradition and hosted by a local group of elders. It was the first time I had experienced the entire ceremony because one had to be invited to participate, and before college, I had not had such connections. Medicinal plants played a role in the ceremony—such as the burning of sweet sage, *copal,* and palo santo. The intensity of the heat bound by the walls and crowded bodies was amplified by the steam that burst forth from the contact of water on the hot rocks at the center of the lodge. In darkness, with my skin pulsing from the heat and drenched in sweat, I could hear the guiding words of the elder, "Reflect on the following..." This began to connect me deeper to a process of cultural rooting as well as with fire and water, the elemental forces of the process.

So even though fire was still serving a utilitarian role in respect to the lighting of sage and *copal,* fire was not simply a tool anymore. It was a cultural element critical to practices of my identity wayfinding.

Years later, in graduate school, my understanding of fire's cultural importance continued to expand, its role slowly unfolding. One of my academic interests was how to support people in environmental and conservation behavioral change. We looked at what motivated people, how our emotions and values play a role, what scared and inspired us, and other aspects that facilitated or impeded people in acting for the environment. I recall my professor sharing research about what calms us and what excites us. In one experiment, researchers wired people up to detect their excitement level and placed them in front of a fire that was more campfire than wildfire. Contrary to a general expectation that we might always react adversely to fire as a survival instinct, the research showed that certain kinds of fire were calming. Part of the theory is that this calming effect may be adaptive, as our early ancestors were more prone to relaxation by a campfire when surrounded by family and immediate community. A campfire was also where they would pass down generational information through storytelling. I think anyone who has gone camping and sat by a campfire can relate and connect to this—feeling the contrast of the warmth of the fire with the outdoor air, the smell of wood burning, the crackling of the fire that can serve as a calming white noise, the feeling of community and conversation with others present. So, we evolved with fire in a variety of ways, not just simply as a tool. We have been shaped with and by fire beyond its being something fearful and destructive, something we seek to control and dominate.

Some of this was new to me, but I would argue that it was certainly not new to humans. Before settler colonization, Indigenous communities in California such as the Yurok, Karuk, Hupa, Miwok, and Chumash actively and intentionally worked with fire to landscape parts of the ecosystem. This kind of fire-based land management was suppressed until recently, to the landscape's detriment, but even though this use of fire was limited, it remains a core component of those communities' relationship with the land. Their knowledge and engagement with natural forces and their observations of natural workings are grounded in being in a relationship with nature as a part of nature, not apart from it. Fire, to them, is not an either-or binary.

In *Braiding Sweetgrass*, Dr. Robin Wall Kimmerer, a botanist and member of the Citizen Potawatomi Nation, shares some lessons about fire that resonate with my own experience of fire. First, recognizing two sides to fires: creation and destruction. They are both there. Next, there are also four types of fires. First is the fire that you make for utility, to cook, keep warm, and "keep the coyotes away." Second is the Thunderbird fire, which comes from lightning strikes but is also the fire that shapes the land. It is possible for it to be highly destructive, but it is also the fire that nature needs. Third is the sacred fire of ceremony and prayer, used for healing and in sweat lodges—a symbol of life and spirit. Fourth, and the hardest fire to care for, is the fire in the heart. She recounts her father sharing: "Your own fire, your spirit. We all carry a piece of that sacred fire within us. We have to honor it and care for it. You are the firekeeper."[2] That sticks with me.

Yet as a colonial framework shaped the development of a place like the United States in the nineteenth and twentieth centuries, we limited fire and stripped away its multifaceted being, ignoring how it can be more than destructively bad or only utilitarian good. The mechanistic, reductive logic that drove colonization led to the development of totalizing approaches to fire suppression by agencies such as the US Forest Service. This approach sought to "protect" the forest for its harvestable wood as a natural resource, compared to supporting its thriving as a community of natural relatives. Through this, we were creating conditions in which our past practices, compounded by climate change, would lead to intensive, destructive wildfires from any spark. Nonetheless, there were still voices calling to reconnect fire to the land—for example, the conservationists A. Starker Leopold and Harold Biswell from the University of California, Berkeley, in the 1940s and 1950s. This eventually led to small changes in the prescribed-burn rules of the National Park Service and US Forest Service in the 1970s. And it was in alignment with ancestral practices of our human and nonhuman kin: considering fire part of the life force of the forest.

In ecology, there are fire-adapted ecosystems in which fire is critical to regenerative cycles. Succession, the process by which a mix of species and habitats in an area changes over time, begins after a fire passes. For example, fire suppression in sequoia forests resulted in a lack of sequoia seedling growth, which led to new prescribed-burn recommendations for the National Park Service in the 1963 "Leopold Report." Low-intensity fire gives sequoias three things for regeneration. First, it provides space for more light and water to be available for young seedlings as "holes" in the forest are opened. Second, the heat opens up mature sequoia cones, releasing the seeds. Last, it clears the underbrush so seeds can land on bare mineral soil and then germinate. Winter comes, burying the seeds in snow, and when spring arrives, they have ideal conditions to become seedlings. Healthy succession is part of healthy regeneration, and fire plays a role in that.

By now, I think we know all too well what happens as we reactively suppress fire in natural landscapes. The absence of fire harms fire-dependent ecosystems and throws things out of balance. This absence also sets up conditions for more disastrous fires, which feed off built-up vegetation. Removing "good fire" aligned with the rhythms and cycles of nature gets us in trouble.

I think of my work in equity and inclusion as being as changing and varied as a landscape. The language of ecology can guide us to become better practitioners of diversity, equity, inclusion, and justice, or DEIJ—which I sometimes refer to as "an ecology of DEIJ."

To start, I have taken diversity, equity, inclusion, and justice and reframed them as decolonize, ecologize, indigenize, and joyify.

To ecologize is to ask, How can we be guided by nature to be radical and revolutionary? Not *radical* as in an external disruptor but as in honoring the root of the word: "to root" and "to be rooted" in the work. *Revolutionary* not simply in inciting dramatic change for the sake of it but also in connecting to revolution as a cycle, being part of regenerative cycles that we find in nature. For it is when we disrupt and destroy roots and regenerative cycles that we find ourselves enmeshed in contemporary environmental challenges and disasters.

In this respect, fire has become a type of mentor to me, showing how sociocultural systems can benefit from periodic and thoughtful regenerative disruption.

Just as we have been reconnecting to ancestral knowledge systems and valuing the role of prescribed burning in ecological systems, I ask myself how that can also apply in our sociocultural systems, especially in white American dominant cultures founded on ideologies of oppression. What prescribed burning is useful in our relational spaces? Within the structural elements of society? To set fire to a space should be practiced with more nurturing and intention than simply a desire to "burn it all down." This is part of being responsible fire tenders. And this may include cultural burning as well, which serves the greater purpose of cultural sustainability.

How can we be as disruptive as fire but connected to revolutions—cycles—that are part of sustaining and even thriving communities? It can be easier said than done because we might still be quite fire-averse, and understandably so. As we have been paying the costs, literally and figuratively, for how we worked against natural cycles of fire, I have noticed the same in our social spaces. If we only suppress fire, then we cannot be surprised when a spark ignites resulting in an intensive, destructive wildfire in our social spaces, and we cannot always firefight our way through it.

We may be prone to resist conflict and tension or engaging in the hard work to tackle the isms—the ideologies of oppression—we've inherited like racism, sexism, ableism, and so forth. When we do, however, we suppress healthy disruption and let the underbrush grow or the forest become unhealthy. In such scenarios, some people may understandably call for it all to burn down, the whole system. Because they cannot untangle what is oppressive and what is regenerative, they determine that the only way to change is for it all to go.

So, I think again of radical and revolutionary change. Rather than destruction, careful "burning" can enable the potential for regeneration—a nurturing burning to see which seeds will grow from such a space. I think of this as work that leads not reductively with a fire of utility and destruction but with a fire of the heart and creation of the kind that Dr. Robin Wall Kimmerer references—a fire that brings life, a fire alive in our cultural and spiritual selves, a fire that is intentionally one of creation.

To be in relationship with fire as responsible fire stewards in both our ecological and our social spaces requires a mind shift along with changes in our practices. I think about how I have reacted to fire with fear and anger, which is how some of us may react to tackling systemic oppression. An alternative is to respond to literal and metaphorical fire with reflection, creating the conditions in which fire will find a home in us not in a way that burns it all down, but in a way that promotes thriving and even healing.

Fire for healing. It may sound odd, and yet, similar to the ways fire can help a landscape heal, I see it also offering healing in our relations through cultural and spiritual roles. From ceremony to metaphorical framing to practice and more, we can burn that which no longer serves us so that we have a landscape for healthy succession, moving from illness to wellness in all our spaces. I think of this when I use the expression *la cultura cura* (culture heals), which helps me reflect on the elements of my Latine culture I want to seed forward and transform with fire, particularly replacing toxic masculinity in the form of machismo with models of positive masculinity that can be threatening to colonial patriarchal structures.

This is an embrace of fire as medicine. This is an embrace of fire as an integral part of healing our severed connections. This is an embrace of fire as a mentor, honoring the ways it can function in the landscape and serve as a model for how "fires" can operate in our sociocultural spaces. I look forward to continuing to learn on my life journey.

notes

1. Erika Buenaflor, *Cleansing Rites of Curanderismo: Limpias Espirituales of Ancient Mesoamerican Shamans* (Rochester, VT: Bear & Co., 2018), 83.
2. Robin Wall Kimmerer, *Braiding Sweetgrass: Indigenous Wisdom, Scientific Knowledge, and the Teachings of Plants* (Minneapolis: Milkweed, 2015), 362–64.

Pyrosketchology: The Language of Fire

Miriam Morrill

The wind can be a persuasive messenger when you speak the language of fire. Speaking that language starts with recognizing that fire is more than just flames and smoke; it is a chemical reaction, a natural force that forms ecosystem processes and forges relationships between the sun and Earth, plants and air, people and places. The language is spoken between discrete and interactive elements in the environment, like the shape of mountains, assemblages of vegetation, and the shifting patterns of winds. The language is learned through observations akin to those of the Jedi master, who senses the seen and unseen forces that influence fire.

I had my first wildfire experience in 1992, working with a small fire-engine crew from the Tahoe National Forest, chasing lightning strikes that had spread across the northern California mountains. The thunderstorm and downdrafting winds had passed over the day before. It was a steep climb through a pine forest to reach the high rock outcroppings at the peak. Red-orange flames waved around the trunk of a cedar tree like the skirts of a flamenco dancer, surrounded by a sixty-foot circle of knee-high blazes that danced in accompaniment to the music of light winds in the branches. Smoke lightly billowed into the air with the scent of burning cedar, like incense in a place of worship. I was struck by the shape of the fire, the circle of flames, like a wedding band, which seemed to symbolize the fidelity and fiery relationship between plants and fire, air and Earth. Not until years later did I learn that fires typically start as a circle that grows outward until there is a change in the slope of the land, the winds, or the concentration of vegetation, which creates new shapes and patterns.

My initial firefighter training emphasized basic fire terminology, as well as fire behavior concepts and tactics, but it was a language of words that I memorized and theorized. I learned about environmental science and ecology in college, but that knowledge was also formed around words and concepts, not the skills of personal observation that allow knowledge to resonate and take form. Over my career, I observed many different wildfires and prescribed fires around the country. I have had fire assignments in Australia, Micronesia, Jamaica, and other places, and the basic fire principles still apply even though each place has its own circumstances. Firefighting and prescribed fire practices around the world are also similar, but each is still unique. I can't help but look back on my education and career and think how unfortunate it was that I did not start with the language of observation so that the basic fire principles would mean something more to me in each new place and fire that I observed.

My last wildfire experience was during the July 2018 Carr Fire, near Redding, California. The fire killed seven people, burned 229,000 acres of land, and destroyed 1,079 homes. My role was to talk with the news media and the public about the wildfire's status and suppression efforts. My knowledge felt useless in this scenario because the fire behavior was beyond anything I had known. This was the first documented occurrence of a fire tornado in US history, with an EF-3 tornado rating and flames spinning four hundred feet high and a thousand feet wide. I didn't directly observe the fire tornado. I was still traveling to the fire and could see only the dark foreboding smoke on the horizon. I did see the aftermath of destruction from the tornado and the fire as I drove around burned neighborhoods to assist people as they returned to the area. I noticed the steel power pylons twisted and mangled on the ground, a tree that was uprooted next to an unburned house, and houses that lay smoking in heaps and piles.

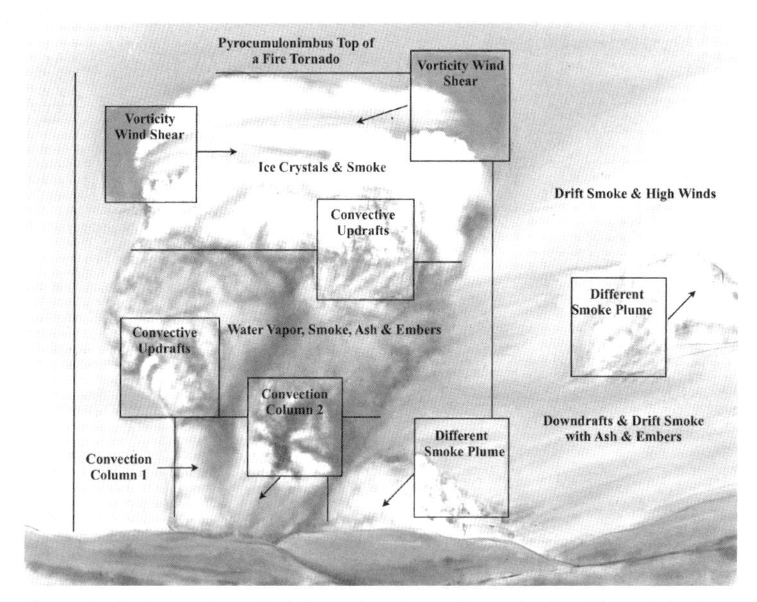

Illustration by Miriam Morrill of the smoke columns that twisted and formed the Carr Fire Tornado. The journaling approach is to sketch the overall shape of the smoke plumes and columns with smaller detailed sections sketched within key or interesting areas to reduce effort and accentuate observations.

My most lasting memory is of being at an evacuation center, when a woman came up to my information board, trying to communicate with me with desperate chirps and clicks while pointing and pounding on the map. I had no idea what she wanted or what to say. Would better visuals have helped? Maybe no information product would be adequate in the face of such extreme fire behavior and with so much loss and trauma. All I felt capable of doing was to witness and be present.

That same year, in November 2018, the Camp Fire started in the steep Feather River Canyon with dry Foehn winds from the east that pushed the fire through the forest communities of Pulga, Concow, Yankee Hill, Magalia, and Paradise in northern California. The fire killed eighty-five people, burned 153,336 acres, and destroyed 18,804 homes and buildings. I watched this fire unfurl from my couch as news images scorched their way into my mind—personal videos of families creeping bumper to bumper with walls of fire on either side of the road, homes swallowed whole by the flames, skeletal vehicles on the road. I grew up in this area, and many of my friends and family members lost their homes. It's terrible to observe a fire from news sources and see only the most horrifying parts of the story and feel the uncertainty and fear that is triggered.

Although the catastrophic qualities of fire are the ones that most people remember, fires are very complex, and our relationships with them even more so. Our relationships with a place can mingle with many layers of emotions, and they often require personalized approaches to healing, learning, and living with fire. The fire season of 2018 traumatized and motivated me to further stretch my skills and explore new modes of fire communication and education.

In this respect, I have used art to find meaning and inspiration in my life, and I have used it for stress management and trauma therapy. When office-bound, however, art began to feel more like an alternative to and an interpretation of the world, a way of keeping the world at a manageable distance rather than a way to become aware and engage with it. As I investigated approaches for building awareness and reconnecting with nature, I discovered the practice of nature journaling: recording multisensory observations in a journal using a mix of narrative descriptions, measurements, and visuals, such as maps, diagrams, and sketches.[1]

I started nature journaling several years ago, taking a small sketchbook on my walks around the neighborhood and a nearby natural area. I had an immediate and powerful reaction to this practice. I noticed details about plants I had previously overlooked, like the hairy, sticky stems of *Madia elegans*, otherwise known as tarweed. When I paid closer attention, I noticed how these plants bloomed throughout the summer, how the vibrant yellow flowers were synchronized with the fire season, and the strong medicinal scent permeating the air on hot summer days. My curiosity piqued,

I researched this plant and found that Indigenous peoples used it as a grain source and burned these open areas to help maintain conditions for the plants. These observations and the journaling practices created deeper connections with that place and greater understanding of human cultural relationships with fire.

The 2018 fire season prompted me to consider how nature journaling might contribute an important and amazing tool to the field of fire education, and the idea offered me a spark of hope that others could cultivate a deeper sense and awareness of fire. Nature journaling may even lead to a heightened sense of awareness of place, an "early alert" system of direct observation that can be used to respond to things happening around us.

In 2019, while still managing a number of California fire-planning and education programs, I coordinated a pilot workshop for a group of adults and young students to observe and journal about a prescribed fire. My partners included the professional artists and nature journalers Robin Lee Carlson and John Muir Laws, the Nature Conservancy's Fire Training Exchange (TREX) program, members of the Karuk and Yurok tribes, the US Forest Service, and the Western Klamath Restoration Partnership. The workshop took place in the rugged Klamath Mountains of northwestern California. Our nature-journaling group took notes during the fire team briefings, during weather and vegetation monitoring activities, and during the fire burning operations. We noticed how the morning test burns would not ignite and spread in the damp leaves and cool morning weather. Later in the morning, we observed how the flames snaked along the path of ignition fluid that had been laid down by fire practitioners in the deep layers of leaves and pine needles of the forest. We watched with fascination as white colored smoke billowed low in the tree branches and light rays streaked between the tree trunks. As the day started to heat up, we saw the flames grow in size and spread faster, igniting shrubs and small trees along the steeper slopes. The smoke lifted above the treetops in a white and congealing plume. Late in the afternoon, small and distinct smoke columns rose into the air where the fire was more active while a larger mass of smoke spread thin and wide across the sky in a brownish tinge.

As we shadowed those engaged in prescribed burning activities, different local experts spoke to us about burning goals, objectives, and tactics. The US Forest Service employee and Karuk descendant Frank Lake showed us the bug-infested acorns in the thick layers of leaves and explained how fire would remove the leaves and infestation, allowing fresh acorns to be collected by the tribe. Margo Robbins, of the Yurok Tribe, brought a selection of tribal baskets and shared knowledge about the plants used for different baskets and how fire helped her people manage forest conditions for the best use of plant materials. Bill Tripp, the director of natural resources and environmental policy for the Karuk Tribe, guided our group to a lookout point along the steep canyon road, where he pointed out historical tribal lands and associated burning activities and practices.

A nature journal page by John Muir Laws capturing a morning briefing before starting prescribed fire ignitions during the 2019 Klamath TREX event.

A nature journal page by Robin Lee Carlson of the observations during the initial fire ignitions during the 2019 Klamath TREX event.

At the end of each day, our journaling group shared observations and journaling techniques, expanding our knowledge and understanding of the observations we had made. On the last day, our group gave a presentation to the fire team and local community and provided electronic photographs of our journal pages to help support local communication and education efforts.

This type of workshop is the ideal opportunity for applying journaling practices, but the foundational observations can be done before a fire and in a variety of locations, even a backyard. Many observations are also made after a fire, when it is safe to explore and observe a burned landscape.

In my pyrosketchology practice, I have framed many of the observations around the key elements of the fire behavior triangle, which includes topography, fuels (vegetation), and weather. The key to these observations is to notice the interactions between the three elements. I also like to frame pyrosketchology observations around the changing conditions that influence fire behavior and what we would call fire hazard levels, which shift over time.[2] In pyrosketchology practice, it is best to include a mix of observations, questions, thoughts, and emotions. When relating to fire, our personal internal environment is just as important as the external fire environment. We react emotionally and bodily to the sheer awe of observing flames, which can also trigger traumatic memories. Recognizing and honoring these emotional reactions to fire observations can help us reframe issues and create less biased responses. Fire expertise or artistic skill is not needed to start journaling or to make astute observations. What is key is a regular practice that engages your senses, captures observations, and makes comparisons in regard to changing conditions over time.

A nature journal page by Marley Peifer capturing the scene of local grade school students journaling about the flames during the Klamath TREX prescribed fire.

Topography

In my writings and workshops, I like to emphasize topography observations that include elevation, aspect, and slope, which influences the weather, vegetation conditions, and fire behavior. In my journal diagrams and sketches of landforms, I add notes about different elevational heights and associated assemblages of vegetation. The aspect or direction the slope is facing (north, east, south, west) will change the level of heating from the sun at different times of the day. By adding notes about the aspect and the amount of sunlight hitting the land, I can key into where and when the land and vegetation are heated, which influences fire behavior. The southern and western slopes typically provide the conditions for higher-intensity fire behaviors.

The angle and slope of the land influence the movements of wind, and thus the careful observer can identify where erratic fire behavior may occur. The shape of the land is something we can observe through the language of our bodies by means of proprioception or kinesthetic senses located in our muscles and joints. At a distant viewpoint, I can see the shape of the land and gain a deeper sense of its slope or steepness by using my hands, stretching my arm out in front of me, laying my thumb at the base or zero angle, and stretching my forefinger to the rise of the slope. These observations can be sketched as simple shapes, with shading and symbols, or both, used to indicate different patches of vegetation. For example, a simple sun symbol can be used, with lines pointing to the slope aspect and notes about the time of day, along with a small triangular diagram with notes about the estimated slope.

Fuels

Fuels are what feed and carry a fire, which often starts in grasses and moves into shrubs and trees. When you observe vegetation as the surface and flow of fire, you can build a sense of fire behavior on the landscape. I like to use simple diagrams, such as a small

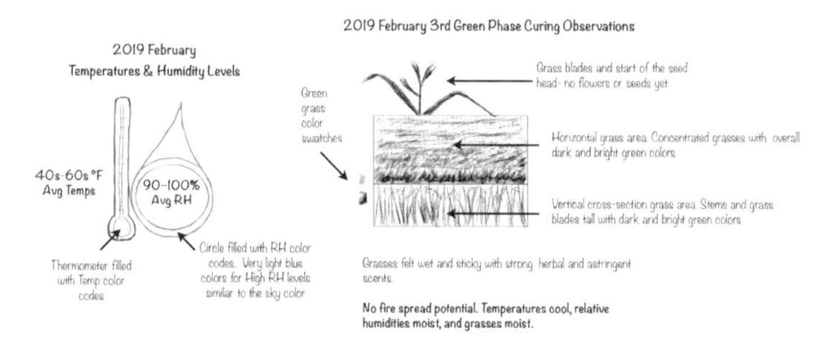

Example of a nature journal page by Miriam Morrill including observations about weather and fuels.

square with a bird's-eye view to convey the horizontal arrangement of vegetation; grasses can be conveyed in colors, shrubs and trees, as circles and triangles. By adding a cross-sectional diagram, with a simple baseline to indicate the slope of the ground and simplified grass, shrubs, and trees, I can convey the potential vertical flow of fire. To engage my other senses, I like to measure my steps between shrubs and trees and use my arms or body to stretch between shrubs and trees to estimate the distances fire may stretch and flow.

It is also important to observe grasses' curing phases over time. Even with a high concentration of grass, a fire will not ignite and spread when most of the grass is green and there is wet weather. Grass curing is a phenological phase. As flowers bloom and seeds develop, the plant loses moisture until the aboveground grass stems, blades, and seed heads become a dry husk and highly flammable. As the percentage of cured grass increases, the potential for fire ignition and spread increases. I use four phases of landscape observation for curing grass: green, green-gold, gold-green, and gold. I use horizontal and vertical (cross-sectional) diagrams for these observations as well. The horizontal view includes a percentage of cured grasses in the area, and the vertical diagram notes the average percentage of curing within the grass stems, blades, and seed heads. I find it best to use a location that has a patch of unmanaged grasses that I can regularly access. As I observe the changes over time, I create the diagrams for each phase, with all of my sensory observations and an estimated percentage of curing.

Weather

Weather observations are my favorite and what I start with in the field. The core weather observations for fire are temperature, humidity, and wind, with hot, dry, and windy conditions elevating the fire conditions. I also love the full-body sensory engagement used to observe the language of weather. It's a gift of awareness that when the wind roars through the trees, the sun scorches my skin, and the dry air burns my nostrils, I can recognize that the fire risk is high. It's a conversation with nature.

Temperature

Journaling about temperature changes is important, especially warming temperatures that influence the grass curing and the flammability of vegetation. Our bodies are fairly poor at recognizing distinct temperatures, but we can sense general ranges, like cool, warm, and hot. Fire can burn in a wide range of temperatures, but the environmental conditions that help it spread are often when the temperatures are warm and hot. I use weather forecasts to practice my sense of temperature shifting, from comfortably warm in the shade to hot, describing how it feels on my skin, in my nose, and down my throat. I also use a color code for those sensed temperature ranges, like an orange color for temperatures between 80°F and 90°F. I use the color in a small box or add the temperature data on top. I can also use these color codes for creating temperature summary graphs, enhancing the meaning and analysis of temperature observations and trends.

Humidity

Relative humidity (RH) is another important weather observation. The drier the air, the more flammable the cured grass and the quicker the fire will spread. Our bodies can sense conditions of high humidity and very low humidity, with sticky or dry skin, but there are other more obvious things we can cue into. Clouds can inform us of air moisture conditions, with lower-hanging layers of clouds indicating higher humidity levels. The sky's color can also inform us of RH levels, with brighter blue skies typical of cold, dry air. I like to use the touch, smell, and sound of cured grass and leaves to observe and relate RH levels. A cured blade of grass, pine needle, or leaf in my hands will crack, break, and or crumble when the RH is very low and the fire hazard is high. That same blade of grass may just bend or twist when the RH is higher and fire hazard is lower. When you integrate weather data with multiple sensory observations, you can fine-tune your observation skills and awareness to the changing weather.

Wind

The wind can speak to us by touching our skin, billowing our clothes, and shaking the leaves and branches. The wind can inform us about the direction a fire may move and the speed of a fire's forward rate of spread. There are some great tools for tuning our observations to wind speeds, such as the Beaufort Wind Force Scale, which uses observable impacts on vegetation and structures. The scale can apply to things as subtle as lifted and rattling leaves to more extreme events, as when trees are pulled from the ground. Another great observation tool is the Griggs-Putnam Wind Deformation Index, which identifies permanent effects to tree shapes on the basis of past and mean wind speeds in an area. These observations can be as subtle as pine needles sweeping to one direction or as obvious as an entire tree bent parallel to the ground.

I like to journal about wind observations by sketching or diagraming a large landscape using simple landform shapes. I can use swirling lines and arrows to indicate where the winds are moving across the landscape. I also add small sketches or silhouettes conveying the degree of tree bend, using multiple wavy lines and arrows to indicate the direction of higher wind speeds. For lower wind-speed observations, I like to sketch the shapes of a few tree leaves and use arrows to indicate the level of lift and twist, with a few swirly lines to indicate movement and notes about the different sounds of wind in the leaves.

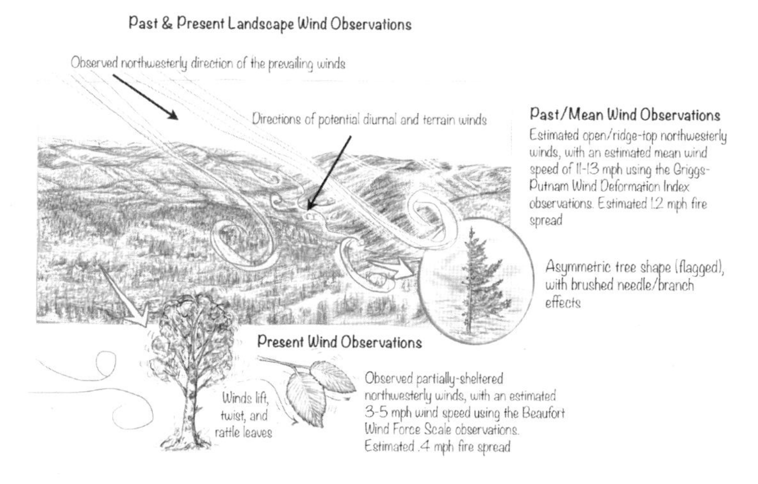

Example of a nature journal page by Miriam Morrill with observations that include topography and winds and how they may influence fire behavior.

Observing and journaling about the fire environment and changing conditions is such a powerful and comforting approach to learning that builds awareness and connections to a place. This is also a tool that can be used to help enhance fire prevention, mitigation, and readiness. By knowing from personal observations when and where fires can start and spread, we can adapt our activities and lives. As a retired person traveling the country, camping in remote areas, I feel a niggling tension about increased fire risk. That is, until I stop and touch the grass, feel the texture, smell the gentle perfume, and listen to the boisterous call of the wind in the trees. Those sensory observations are soothing and help me focus on the moment, settling my nerves. I use nature observations and journaling practices to keep a calm and balanced perspective about fire and to focus my interactions with the land. The language of fire is a wordless conversation with nature.

When you are attuned to the elements and interactions between weather, topography, and fuels, you are learning the language of fire. These ongoing observations and journaling practices can both soothe and heighten your senses and awareness. In a world where technology distracts us and temperature-regulated rooms disguise the changes in nature, the simple act of observation can bring us closer to the living world. A more fire-adapted culture is one where we respect, care for, and perhaps create a greater willingness to listen to and work with fire.

notes

1. For an excellent example of this practice, see "Nature Stewardship through Science, Education, and Art," John Muir Laws, https://johnmuirlaws.com/.
2. More can be found about fire-hazard-level observations, among other pyrosketchology techniques, at Miriam Morrill, "Pyrosketchology—An Illustrated Guide to Nature Journaling about the Fire Environment," Pyrosketchology, https://www.pyrosketchology.com/.

Indigenous Fire

Ron W. Goode

Mun-a-hu e-boso, Tah-qwa Nium, Hunit-to-wah-eu? Eu-hun-sut chow-ute?
Neu Tie-now-wah Togu-pah-too, ah-gee-yah eu to-bohp, pi-yah, kos, wa-be sah-bee!

A long time ago, when the world was being made, Water (Pi-yah) married Land (To-bohp), and they had a mischievous child, Fire (Kos), who was always playful and loved to leave fire everywhere. Water followed him around, sprinkling his trail, then Land followed behind, covering up his tracks with flowers and fresh resources.

What we as humans know is that our bodies are made of Earth's elements and that the majority of our body is water. We also know that our blood is our soul; it is where our spirit lives. Without blood, no spirit—no life. With Earth and Water intact, our first tool, the first offspring, was Fire. Fire is our passion; without Fire there is no passion.

We know that after a wildfire, the water returns; then Mother Earth starts growing new trees, new plants, beautiful flowers. Our creation stories tell us about different animals and birds stealing Fire from the Sun, others taking it away from them to protect everyone. These stories tell about those who sought the power of Sun, controlling it, misusing it, abusing it. Not the light, but the heat and the brightness of the white light—the making of Fire.

The Nium (North Fork Mono) creation stories are passed down generation after generation, telling us when our people lived in the metaphysical world. When the world was being made by Tie-now-wah Togu-pah-too (Creator), our people were animals, birds, insects, and reptiles. We emerged and transitioned from the metaphysical to the physical world, bringing with us the spirits of our relatives; our "pets" or our "puk," as the Nium refer to them, are still a part of us. We know that Creator gave life to all beings; therefore, all life is in relation. In the metaphysical world, all peoples spoke to one another. They spoke many languages, but the language they spoke the best was the language of ecology.

Fire is a tool the Indigenous people use to ensure the betterment of their livelihood—a tool used for millennia to take care of their landscape, their resources. Fire, like water, is sacred because the plant resources and land calls fire, fire calls wind, and wind brings the water. They all work together, all important to the welfare of one another, important to the sustainability of each one's health.

If fire is something you are afraid of, or if just the thought of it creates anxiety, then it will not be sacred or a tool. Extreme wildfires are a result of ill-mannered, mismanaged profiteering by people who live by and adhere to the concept of commodification. Extreme wildfires result from decades and centuries of economy-first thinking, from commodity-based concepts that created suppression, oppression, and repression. Bounties were put on Indigenous burners' heads for lighting fires and keeping their resources refreshed.

When Euro-Americans arrived in California between 1800 and 1850, the trails were open, the landscape provided easy traveling, and bountiful resources of fiber, food, and medicine existed everywhere. Pre-1850, when Native Americans were living on these lands, they used fire to keep the overhead canopy at 40 percent thickness or less. Fast-forward 170 years: the mega-wildfires, devastating drought, and bark beetles have had their way with a forest and foothill canopy that has 80 percent to 90 percent thickness. Five large fires hit the Sierra National Forest before the Creek Fire of 2020, which burned 378,000 acres. Two-thirds of the Sierra National Forest's 1.2 million acres have burned over the past eight years, and still the Forest Service has done little to ensure a more resilient landscape over the remaining one-third of forest.

Photos of the Yosemite Valley from 1850 to 2022 all show the degradation of the valley based on the commodity thinking of early settlers. An 1872 photo shows an openness of the valley floor. At this time, sheep and goat herders moved in. These early settlers blasted the meadows so they would drain faster, allowing their stock to have better grazing. That concept changed the ecological ambiance and allowed conifer encroachment.

By 1927, the valley floor was fully covered but not thick with vegetation. A 1950s photo shows even the walls of the great valley covered with trees. That progression of time is relevant to what occurred to the rest of the forest as well. In 2023, even the meadow on the floor of Yosemite has receded. The snapshot of this iconic "natural wonder" is the epitome of the forest north and south of Yosemite National Park, down to Sequoia National Park. Of course, since 2020, large wildfires have reduced those national forests and the forests of our national parks to rock and black candlesticks.

The devastation of our forest lands has created a chain of reactions. As it is said, "Where there is an action, there is a reaction!" With 378,000 acres of trees demolished by wildfire, that's millions of trees no longer sucking up water. The North Fork Mono Tribe assessed fifteen meadows following the Creek Fire of 2020; every meadow was full of water, soaking boots and pant legs up to the knees. That would be a positive outcome except all the meadows have huge gullies, and the bottoms or spillways are blasted. A meadow's ecosystem is supposed to absorb the water, holding it and then releasing it downstream. Now, with the fast runoff, the waterways in the meadows are deepened, and the meadow does not fully function as a sponge. Instead, the water runs off the meadow, leaving it to dry out once the warmer temperatures arrive, drying out the surrounding watershed that is already stricken by drought and bark beetle. With trees in place on the hill and mountaintop, normally the Mono Winds blow at 60 to 70 miles per hour. In 2021 the winds blew at 90 miles per hour, taking down large conifers and oaks whose roots and base soil had weakened and were easily toppled. The Mono Winds of 2022 were recorded at 110 miles per hour. Remaining snags left by the Creek Fire and other previous fires snapped in half, reducing any sort of a buffer that might slow the wind down.

The rainy season of 2022–2023 in central California produced some of the heaviest rain and snowpack totals—matching previous highs in its history—and at the time of this writing, the snowpack was recorded at 191 percent, with another atmospheric river on its way in. With no buffer or trees in place, there is nothing to hold back the runoff. With warm rains, flooding has already begun. Not only is California experiencing the chaos of climate change, including our thirty-sixth year of drought since 1987, the anomaly of the atmospheric river mixed with mismanagement and colonial economic philosophy is a prescription for disaster and exacerbates the impacts of climate change.

How, then, do we build resiliency around government agencies and leaders unwilling to change from "management" to "stewardship"?

It's not that agency folks don't understand the difference between management and stewardship. They do. Colonial thinking comes not only from within the agency but from environmentalists as well. When asking agency folks why they burn or what they burn for, acreage is the answer they give. Acreage, because that's what they get paid for. Therefore, economics is what the game is all about. Environmentalists want larger burns, more acreage. That's the concept put forth in all strategies or in official strategic plans. This game of economics will play out for decades to come. The Forest Plan Revision, which has been in draft form for thirteen years and will be enacted for the next twenty to thirty years, has two game plans: recreation sustainability and economic sustainability. Why no forest health sustainability? So the agencies can remain flexible in their decision-making. Recreation is the number-one economic endeavor. Economic sustainability means logging and conifer products. Therefore, economics is both one and two—and that's it, folks!

When various state and federal forest-fire crews burn, that's what they are burning for—better tree production. These crews eliminate the undergrowth; they eliminate what feeds or provides the nutrient to the trees. A current example of this thinking is in a new emergency order protecting the sequoia tree groves. The "order" is to scrape five to ten feet around the base of the giant tree, then remove plants and shrubs within a twenty-five foot radius of the tree and lower conifers within one hundred feet. Removing or scaping the surface around the tree trunk means baring the root of the tree. The root system is at the surface; therefore, when scraping, the root will be scraped and damaged. Removing the plant life from the floor twenty-five feet out means removing the food supply to the tree. Gooseberries, blueberries, ferns, and other vegetation are not going to light the tree up if they burn. A fire will actually restore them, refresh them, and provide more nutrients for the root system. Clearing the taller conifers will help, but what will really help is to make sure all the dry-ladder fuel is removed so a fire will not climb into the conifer treetops and allow the fire to jump to the crown of the sequoia tree.

Here's an analogy of this colonial thinking: You are invited to dinner at your friend's house, you enter, and the food already smells so good. A large table awaits you, with really nice plates and silverware set out for you and your family. Then the cook comes out, picks up all the plates, silverware, and glasses, hauls them back to the kitchen, and closes the door so you can no longer smell the dinner. Then the host asks you how dinner was: Did you enjoy it? Well, that is exactly what is being done to the sequoia trees. All their food is being removed.

The Indigenous burners, the cultural burners, want a return of the undergrowth of the resources they are burning. The purpose of cultural burning is to get that return. There is an expectation for an immediate return. The fires are not so hot that the root system is killed, as in a severe wildfire or as hot as an agency-prescribed burn. With cultural fires, that is, "good fire," the plants, shrubs, flowers, and medicine plants have an expected return of one to two months. When they do grow again, they are refreshed and rejuvenated; the leaves, stems, branches, and roots retain water and hold water for longer at the surface. Thus, the water table rises after a fire, but after a good fire, it is held longer at the surface.

The difference between cultural fire and a prescribed fire is in the size of the fire. Cultural burners make small beaver-hut-sized fires. No one sees or complains about cultural fires, not even ranchers' fires. Ranchers have burns going on in foothill-mountain communities quite often, sometimes six to seven fires going in every community—and that's considered normal smoke. Not so with agency fires; those affect everyone near the fires and down in the valley as well. Folks with health conditions in the valley are immediately affected by the hundreds of acres being burned. Agency and air pollution folks can tell you by name who these people are because they are affected every time.[1] The particulates from an agency burn or a wildfire create an immediate reaction. On a broader scale, the smoke and fire particulates affect not only the people but also their homes, their insurance, and their overall health. Law firms send out notices: *Your home was in a fire zone. Your home was recently affected by such-and-such burn. You need to file a law suit or join this settlement.* If your social well-being is affected because you were actually in or near a large fire, try getting insurance; many insurance companies will no longer insure homes in a "fire zone." However, good fire and good smoke help to improve the health of the forest and the habitat, for the wildlife and for the natural resources.

Indigenous burning was constant, consistent, and practiced for thousands of years before colonization. The North Fork Mono

Tribe and tribal team have conducted numerous burns over the past twenty years, burning on national park lands, in national forests, on tribal lands, and on state and private lands. All these small-scale burning practices are designed to improve the cultural resources and the overall health of the landscape.

In addition, the tribe conducts studies on production of acorn products, creating a higher quality and quantity of basketry material, Native food sovereignty, and medicine for the people and the animals. These studies have shown a large increase in production. Cultural fire in the Sequoia National Park over a three-year period indicated an increase in the 215 monitored acorn-producing blue-oak trees from 36 percent to 54 percent. On a meadow in the Sierra National Forest, after cultural burns, the increase in acorn harvesting rose from twenty-five pounds per producing black oak to forty and fifty pounds of acorn per tree.

There is a place and space for both cultural fire and prescribed fire, for both "burners." The same goes for both stewardship and management. Stewardship is overall care for the land. Management is the decision aspect, but management decisions should be based on long-range stewardship of the land, not only on current economic needs. In this manner, decisions on good stewardship will result in a healthier forest. Cultural-style burning creates a "defensible" landscape as well as a more vibrant forest.

This essay began with Nium (North Fork Mono) creation stories, describing how the landscape was abundant with resources, and then the Euro-American arrived bringing "commodity thinking." Let's end with "The Hummingbird Story," a story told by different tribes in their own way. This is the way the North Fork Mono tell it: One day there was a big fire in the forest. All the animals fled in terror. They went in all directions, because it was a very bad fire. Suddenly, We-hay-shet the Cougar saw Pish-ca-tu the Hummingbird pass over him. He was going in the opposite direction, right toward the fire!

Whatever happened, he wouldn't stop. Moments later, Cougar saw him pass by again, this time in the same direction as Cougar was going. We-hay-shet observed this coming and going until he decided to ask Pish-ca-tu what was he doing. It was pretty crazy behavior.

"What are you doing, Hummingbird?" Cougar asked. "I'm going to the lake," he answered. "I drink water with my beak and throw it on the fire to put it out." We-hay-shet laughed: "Are you crazy? Do you really think that you can put out that big fire on your own with your very small beak?"

"No!" said Hummingbird. "I know I can't. But the forest is my home. It feeds me; it shelters my family and me. I am very grateful for that. And I help the forest grow by pollinating its flowers and plants. I am a part of her and the forest is a part of me. I know I can't put out the fire, but I must do my part."

At that moment, the Forest Spirits who were listening to little Hummingbird were moved by his loyalty to the forest. And then, miraculously, they sent a huge downpour of rain, which put an end to the big fire.

Tribal elders tell this story to their grandchildren. Then finish with "Do you want to attract miracles into your life? Then do your part!"

notes

1. The State of California is divided up into districts, which are overseen by air pollution boards. Boards comprise political leaders, professionals and business leaders, and highly concerned citizens, who decide policies for their districts on air pollution.

Descent

Rina Garcia Chua

August 2021

The past two days, ash fell
in Kelowna. Small pieces,
really. Grey smoulders that
disintegrate when pinched
in between my fingers.

It is everywhere: on the dash
of my car, crown of my head,
all over my crisp black shirt;
feeding the dying garden
a whiteness, undigested.

North from where I am, a burning
intoxicates neighbouring cities.
Some people refuse to move
from the only homes they've
ever owned—lands not theirs.

I've packed all my essentials
in one easy bag to shove into
the car if evacuation orders
extend into the city. Fires burn
both ends; no way out again.

As a child, I watched as ash
fell from the unsettled skies
of Manila. I wrote my name
on ash piles atop cracked
cement. I moved dust and

the strange, new grey
chalky matter into corners,
thinking, Is this what snow
looks like? How do people
enjoy what seems so perilous?

Dad rushed out to force
me back inside the house;
Pinatubo has erupted, he
gasped, turning the radio
dial way up so we hear.

Years later, we find grey
dust in the nooks of our
furniture; the dials of
the radio. In closets,
inside Mom's precious
vases. In my locked diary.

There is no escaping
when life is bookended by
the burning of elements:
In a myth from my country,
the sun burned the stars—

The moon, upset at their
deaths, ran away from him.
He begged for her return,
but she kept evading him,
constantly on the other side

of the earth. They chase
each other, daily, in a cycle
of denial—running away
until maybe one will falter,
until the skies clear again.

I've gone to the other side
of this earth and tasted
snow. I don't run away
from a torching; fires are
everywhere. We stay

put until the next spark—
the next slash—and I
wonder if there is a life
where the skies stay clear
like a newborn's eyes.

The Burn Test

Nina Mingya Powles

In the garden, under the apple tree, I arrange pieces of fabric on my lap. One is a metallic pale-pink brocade, embossed with gold and bronze blossoms. The other is a colorful batik cotton with intricate layers of floral patterns against pink and turquoise dyed cloth. One of the fabrics is quite weighty; it holds some structure when I twist and fold it. The other is very lightweight, the cotton threads unraveling in my fingers, weighing nothing.

I light a match and the strip of brocade curls away from the flame, then seems to melt into a plasticky residue. There's an acrid chemical smell, an indicator that this is polyester—pure plastic—a fact I'd already suspected. I blow out the match and light another. The batik cotton sizzles and catches alight. The burning produces a papery smell.

The burn test is one way of determining the fiber content of a piece of fabric. I started teaching myself how to sew in 2020, and since then, I have often rifled through piles of unmarked fabrics at charity shops and market stalls. I learned about the burn test, along with many other sewing tips, through a lively global network of home sewists on Instagram. The test dictates that if the singed fabric smells like burning hair, it's likely an animal fiber—wool or silk. A chemical smell and any sign of melting suggest polyester, and cotton or viscose (otherwise known as rayon) smell like burning grass or paper.

Polyester and nylon are made of plastic compounds derived from petroleum. Like most modern fabric-manufacturing processes, whether natural or plastic fiber, the making of polyester is heat and energy intensive. Small plastic beads around the size of a grain of rice (the same measure I use for hand stitches) are melted, and the molten plastic is poured through spinnerettes. These machines spin soft fibrous strands in varying weights, which are then dyed with chemical dyes and woven into textiles.

Fire and energy transform these raw materials—cotton, flax, plastic beads, a silk moth larva's small cocoon—into soft, pliable textiles that we wear each and every day. It feels like a strange kind of elemental symmetry that burning can reveal clues about a fiber's hidden raw components. Fire can't reverse the chemical process, but it can show us what's inside. Charred linen smells like a scrub fire. Hot plastic melts back into liquid.

I can't shake the feeling that the fibers I wear close to my skin are an indelible part of me. I think back through the fabric composition of my life. My childhood was the texture of OshKosh corduroy, glitter tulle tutus, brightly colored rain jackets, and waterproof snow pants. When I started at an all-girls school: thick blue cardigans, heavy pleated skirts, and itchy woolen tights.

My teenage years of the mid-2000s were made of polyester. One of the first items I remember coveting was a cropped fake-leather jacket, soft and shiny, with silver zips on the breast pockets and up to the bomber-style collar. This was during the cursed phase of the ubiquitous fake-sheepskin Ugg boots, which I also saved up my pocket money for. When we moved to China when I was a teenager and I first experienced a co-ed international high school with no uniform, skinny jeans and stretchy H&M graphic tees became my new kind of uniform.

But when I think of my most worn and treasured garments of this time, I think of denim and cotton canvas. I wore red Converse sneakers every day and painted my nails purple or black, my iPod Mini usually clutched in my hand. I remember a favorite jacket, bought for me by my mother in a Levi's sale at a mall next to Jing'an Temple in Shanghai's bustling shopping district. It was cropped just above the hips and made of thick green canvas, with lots of pockets and, unusually, a soft cotton-jersey lining. While wearing it, I always felt like a stronger and tougher version of myself. I kept wearing it even after the pocket seams fell apart.

After three years in Shanghai, we moved back to New Zealand, and it was at that point that I lost the thread of the jacket's presence in my teenage life. Either it went straight to the back of the cupboard, or more of its seams disintegrated, or it got lost somehow in the move.

Now, in my late twenties, I make and seek out mostly linen and cotton garments for their comfortable, hard-wearing qualities (for dog walking and gardening) and for their depth of color. But as much as I try, I can't avoid nylon. Originally developed as a cheap alternative to silk, nylon is finer and more lustrous than polyester, and it's elastic. Nylon gives stretch to underwear, socks, and swimsuits; all my base layers for deep winter, including some I sewed myself, are merino wool blended with nylon. The soft and stretchy fibers cling to our skin, keeping us warm.

The creamy overcast sky of Luoyang felt unbearably bright, even through smog and cloud. We gathered outside the Museum of Intangible Cultural Heritage with a single guide assigned to a group of around a hundred language students. I couldn't make out the second group of characters on the sign above the entrance; the English translation was charmingly vague. Luoyang, a city in western Anhui Province, seemed to me a gray and unremarkable industrial hub, its rich history obscured by factories and shopping malls. But as in most provincial Chinese cities, layers of history and culture did come to the surface at such dedicated sites. And even as a visitor just passing through, at night I could feel the city's light and color in the bustling open-air food markets and karaoke clubs, aunties dancing in formation in the square just as they did in Shanghai.

In the multistory building that felt part shopping mall, part arts and crafts market, dozens of skilled artisans sat in individual studio-cubicles surrounded by examples of their craft. Musical instrument makers, wood carvers, glass sculptors, jade carvers, weavers, embroiderers. The makers were focused on their work and didn't speak to one another—or at least not during our visit—but were content to answer questions from language students who felt confident enough to practice their Mandarin. I was too shy to chat, but I stood for a long time near the studio space of a silk embroiderer who worked shimmering threads through her fingers and through a screen of sheer fabric to create a detailed picture of peonies with unfurling silver-pink petals. I was transfixed by the movement of her hands, the way each turn of her finger left a stitch of color, until the embroiderer paused to drink from her tea thermos and suddenly the unease of this beautiful but strange space caught up with me. On these university-organized trips with their standard tour-group itineraries, I'd learned that sometimes tourist attractions could surprise and disturb. You become a little desensitized to surreal and unexpected scenes, such as real craftspeople and their handiwork being turned into living, breathing museum exhibits.

I thought of the hundreds of carved stone Buddhas we had seen on our excursion to the ancient Longmen caves earlier that morning, some tiny and some monumental, all shaped and carved and placed there by hand.

These attempts to write a personal history of polyester are inseparable from the years I spent living in China with my parents when I was a child and, later, on my own as a university language student. In her book *Worn*, the American historian Sofie Thanhauser traces China's history as a producer of the finest silk and cottons to the world's leading polyester manufacturer.[1] I had hardly noticed at the time, but textile crafts and manufacturing often punctuated my trips with my parents to regional Chinese cities. As a teenager I visited the National Silk Museum in Hangzhou, where intricate embroidered silk panels lined the pale-green halls. On a school trip in seventh grade, we went to a cashmere factory showroom in Inner Mongolia where glossy pictures of goats and goatherds were framed above displays of pastel-colored scarves, socks, and sweaters.

My mother often brought me to an old textiles shop called the Shanghai Blue Nankeen Factory. It was down a long alleyway off Changle Lu, wedged between trendy pop-up bars and cafés. Here, a small team of artisans crafted traditional hand-dyed indigo cloth. The distinctive rustic pattern of white flowers against dark blue became part of the fabric of our home, which was filled with things my mother had collected from all the places she'd lived. Years later, I went back several times on my own and took pictures of the wide bolts of cotton hanging in the sun in the courtyard like blue sails. It was one of the city's many strange contradictions that this studio existed while, only a few streets away, my friends and I had lined up for the opening of China's first flagship H&M store in 2007.

At that time, the heavy smog that clung to the building tops of Shanghai, especially throughout the winter, often smelled faintly like something burning in the distance. On particularly polluted days, when the air-quality app on my phone flashed a dark red warning, I wore a face mask and went out only for essentials. I'd come home with black dust beneath my fingernails and in the creases of my eyelids. The tiny particles must have seeped into my clothing, too. I thought of the beautiful bolts of cloth at the Blue Nankeen Factory hanging in the courtyard between the lane houses, their woven cotton threads that light and liquid seep through, absorbing the dust and smoke of the city. I thought of the way we wear garments to protect our bodies against the elements a rainproof coat, a woolen hat, a fabric face mask. Something about this city made my skin feel porous. The dense humidity in summer, the thick smoke-tinged mist in winter.

On the sixteen-hour night train from Luoyang back to Shanghai with my classmates, I lay in the top bunk in the glow of flickering light through gauzy curtains. Every few hours through the night, a snack trolley rattled past bearing Styrofoam cups of freshly cut fruit, instant noodles, and packets of meat jerky. Jolted awake by the train's juddering movements, I turned toward the window and glimpsed what looked like a burning ring of fire in the middle of a field. I saw it—or I think I saw it—then it was gone. In the days afterward, I couldn't stop seeing it. A huge circle blazing bright orange in the dark.

Plastic is burning everywhere. Nowadays, this particular sentence repeats in my mind. I'm not sure if it's mine or if I first heard it elsewhere. I still see the ring of fire, too, even though I don't know what I saw; it could have been a factory or a stretch of farmland, or I could have been dreaming still. I think of it now in the wake of the catastrophe still unfolding in Palestine, Ohio, where in February 2023 a train carrying petrochemicals, including vinyl chloride, one of the core compounds used to make PVC plastic, derailed and then burned for days.

Crafting a garment, whether made by a home sewist or a factory worker, is a slow process that uses several key elements: the physical handwork of pinning and tacking seams in place, the small but powerful engine of the modern domestic sewing machine, and the heat and steam of the iron. Too much hot steam on polyester fabric will cause it to warp and melt.

I taught myself how to use a sewing machine in the late spring of 2020. I read and reread pattern instructions, pricked myself with sewing needles, and burnt my skin on the hot iron. I sewed garments backward and inside out and cut up old clothes to practice with the scraps. Sewing became a meditative and creative practice for me: a messy, safe, slow space to make mistakes and learn by doing.

For the longest time, I never sought out information about where my clothes were made. I didn't necessarily care to know the answer. In 2019, the first reports emerged of fashion brands sourcing cotton from Xinjiang in China's far west.[2] The BBC initially named Uniqlo and Muji as two leading brands connected to Xinjiang's cotton industry, and I looked at my closet: those were two of my go-to stores at the time. Although fast-fashion companies had been using factories with dangerous labor conditions for many decades, it was only then that I decided to give up on the brands that had shaped those early years of discovering my personal style. This choice of mine was intertwined with financial privilege; newfound stability meant I could for the first time afford the time and cost of seeking out more ethically made garments that I couldn't before.

Learning to sew taught me that all clothing is handmade. I can no longer separate a garment's seams from the hands that stitched them. Sewing also made me realize that plastic is everywhere in the garments we wear, sometimes hidden, sometimes not. When I touch a piece of fabric now, I know that the fibers hold the memory of all the energy and heat that went into the chemical processes of its making. I wish I could say I had known this all along, ever since I first started buying and borrowing clothes, but it's only in the past few years that I've sought to grasp what it means to buy, wear, and accumulate synthetic textiles. It's only recently that I started reading every garment label.

Home sewing is not accessible to everyone, and it's no antidote to the destruction of the fast-fashion industry. Even while I avoid non-biodegradable fabrics, I'm still consuming, still creating new garments that will likely exist on this planet beyond my lifetime. According to data published by Close the Loop, pure linen fibers can decompose in as little as two weeks; cotton takes up to five months.[3] But this isn't taking into account thread, zips, buttons, and other components. Polyester thread is the strongest, most durable thread for sewing; cotton thread is liable to leave residue that wrecks the insides of sewing machines. The truth is that even if the pieces of my homemade garments may biodegrade, the sewn threads will remain.

"The most sustainable garment is one that's already in your wardrobe" is a phrase that's frequently circulated on posts by Instagram home sewists. At first, I sewed as fast as I could, eager to practice my new skill and create garments personalized to fit my plus-sized body that I couldn't easily find elsewhere. But I had to slow down: the impact of work and health got in the way of making. A slower, more thoughtful mode of sewing is what I strive for now.

It was a friend who first encouraged me to buy a secondhand sewing machine on eBay, and since then, three more of my friends have taken up the craft. We swap patterns and fabrics and sewing tips. Whenever we can find the time, we mend, rework, and recut old garments to extend their lives.

The first wearable garment I made was made from batik cotton my mother sent me. She doesn't have a sewing machine, but she stitches things by hand and collects textiles. This lightweight cloth, bought by her on a trip to Indonesia, was dark blue with a resist-dyed pattern of red- and white-petaled flowers. The crimson petals were bright against the dark indigo background, like small floating flames. With it, I made a simple summer shirt that felt cool and loose against my skin. The batik pattern reminded me of a patch my mother stitched onto a quilt made by my grandfather, cut from an old sarong, to cover a hole in the quilt. She mended it by hand, stitching diagonally with white thread across the edge of the fold.

Batik is a traditional craft originating in Java, Indonesia, which then spread to other parts of Southeast Asia, China, and Africa (many African wax prints originally drew inspiration from Indonesian batik cloth that Dutch colonialists imported in the late nineteenth century). Brightly colored Malaysian and Singaporean batik patterns in particular feel like a part of my childhood, with their pink and blue floral swirls and dotted leaves.

On a warm and bright spring day in London, I watch the designer and artisan Yiran Duan melt a small pot of wax on a hot plate. She is demonstrating a batik dye technique that comes from the Miao people of southern China and Southeast Asia. Yiran herself is from Yunnan, where her family runs an indigo farm and where she learned traditional indigo-dyeing methods from her grandparents when she was a child.

As the wax melts, I begin to feel heat emanating from the pot in the center of the workshop table. For the wax to adhere properly to the cotton fabric we are using, it needs to be heated to 150 degrees Celsius, but traditionally the wax would be melted over a large open fire. Yiran tells us that for craftspeople in small villages, in winter months, the open fire doubles as a communal source of warmth, a place to gather.

The wax turns to liquid gold. Yiran dips the head of her wax knife, a special handmade tool with a bamboo handle, into the liquid. This process of heating the metal by holding it submerged in wax is called "opening the blade." When small trails of bubbles appear after a few moments, the knife is ready to use.

I practice trailing uneven wavy lines of wax on newspaper first, getting used to holding the knife. Yiran speaks about becoming familiar with the weight and angle of the blade on the cloth. I'm wary of the heat of the knife at first, instinctively holding my fingers away from it, but I soon realize that in order to have control, I need to move my grip closer to the hot metal. From a young age, we learn to keep away from fire, to not get too close to a hot stove. But proximity to fire and the wielding of intense heat is also core to many crafts and ancient methods of making: the potter's kiln, the glassblower's torch, the bread maker's oven.

We dye our pieces of cloth in Yiran's three-year-old indigo vat, which she keeps in a bucket in the courtyard of the studio. Indigo needs heat to produce deep color, so Yiran adds hot water to the vat before each dye session. Standing in the April sun, we dip our cloth in the dark, opaque dye, holding it there for ten seconds. It emerges a greenish hue before slowly turning blue in the open air. This moment when the dyed cloth comes into contact with the air is called oxidation, when oxygen transforms the organic compounds of the dye into their final true-blue form. Each dip—and each exposure to oxygen—creates another bright layer of blue. I dip my piece nine times in total, searching for a deeper hue.

Yiran swirls the cloth in a bucket of boiling water to dissolve the wax. We stand around her, steam filling our faces, transfixed as white patterns and lines emerge where the wax has resisted the dye. I rinse the fabric under the cold tap and return it to the boiling water several times to melt off the wax, my fingers and fingernails now blue. I wring the steaming cloth in my hands and lay it to dry on the table outside, my handwritten words emerging sharp and clear, the blue deepening in the warm sun.

notes

1. Sofie Thanhauser, *Worn: A People's History of Clothing* (London: Penguin Books, 2022).
2. Ana Nicolaci da Costa, "Xinjiang Cotton Sparks Concern over 'Forced Labour' Claims," *BBC News*, November 13, 2019, https://www.bbc.co.uk/news/business-50312010.
3. "Strategies for End of Life," Close the Loop, https://www.close-the-loop.be/en/phase/3/end-of-life#:~:text=LINEN,the%20fabric%20into%20small%20pieces.

Let Me

Camille T. Dungy

Let me tell you, America, this one last thing.
I will never be finished dreaming about you.
I had a lover once. If you could call him that.
I drove to his apartment in a faraway town,
like the lost bear who wandered to our cul-de-sac
that summer smoke from the burning mountain
altered our air. I don't know what became of her.
I drove to so many apartments in the day.
America, this is really the very last thing.
He'd stocked up, for our weekend together,
on food he knew I would like. Vegetarian
pad Thai, some black-bean-and-sweet-potato chili,
coconut ice cream, a bag of caramel popcorn.
Loads of Malbec. He wanted to make me happy,
but he drank until I would have been a fool
not to be afraid. I'd been drinking plenty, too.
It was too late to drive myself anywhere safe.
I watched him finger a brick as if to throw it
at my head. Maybe that's a metaphor. Maybe
that's what happened. America, sometimes it's hard
to tell the difference with you. All I could do
was lock myself inside his small bedroom. I pushed

a chest against the door and listened as he threw
his body at the wood. Listened as he tore apart
the pillow I had sewn him. He'd been good to me,
but this was like waiting for the walls to ignite.
You've heard that, America? In a firestorm
some houses burn from the inside out. An ember
caught in the eaves, wormed through the chinking, will flare up
in the insulation, on the frame, until everything
in the house succumbs to the blaze. In the morning,
I found him on the couch. Legs too long, arms spilling
to the carpet, knuckles bruised in the same pattern
as a hole in the drywall. Every wine bottle
empty. Each container of food opened, eaten,
or destroyed. "I didn't want you to have this,"
he whispered. If he could not consume my body,
the food he'd given me to eat would have to do.
Have you ever seen a person walk through the ruins
of a burnt-out home? Please believe me, I am not
making light of such suffering, America.
Maybe the dream I still can't get over is that,
so far, I have made it out alive.

Stolen Object or Gift from the Mother? Mythic Meanings for a World on Fire

Jane Caputi

Prometheus... stole the far-seen gleam of unwearying fire.

—Hesiod, *Theogony*, Greek sacred myth

Take this fire as a gift to your people. Honor this fire as you honor me.

—The Fire Goddess Mahuika to the demigod Maui, Māori sacred myth, in Wiremu Grace's "How Maui Brought Fire to the World"

I imagine if we acknowledged that everything we consume is the gift of Mother Earth, we would take better care of what we are given. Mistreating a gift has emotional and ethical gravity as well as ecological resonance.

—Robin Wall Kimmerer, "Robin Wall Kimmerer on the Gifts of Mother Earth," *Emergence Magazine*

In his classic 1949 essay "Thinking Like a Mountain," Aldo Leopold tells a momentous story with profound ecological resonance, one that turns on the element of fire. He and a few other men were hunting on an Arizona mountain when they encountered an old mother wolf with several pups. Seeing only a despised species with whom they competed for prey, the men opened fire, leaving the wolves dying or injured. They went to inspect the mother: "We reached the old wolf in time to watch a fierce green fire dying in her eyes. I realized then, and have known ever since, that there was something new to me in those eyes—something known only to her and to the mountain. I was young then, and full of trigger-itch; I thought that because fewer wolves meant more deer, that no wolves would mean hunters' paradise. But after seeing the green fire die, I sensed that neither the wolf nor the mountain agreed with such a view."[1]

Leopold continues with an instruction in ecology. As the United States extirpates wolves, deer herds grow unchecked, eating all the plants on the mountains, leaving the herds to starve, and the mountains now without vegetation to erode, culminating in "the future washing away into the sea." To "think like a mountain" is to perceive all these interrelationships and act wisely to preserve and promote continuance.

In an earlier essay, Leopold writes of the need to "cultivate... respect for mother-earth, the lack of which is, to me, the outstanding attribute of the machine-age."[2] The old mother wolf with the virid flame in her eye stands for Mother Earth, and the green fire for her gift of the life force–source. This gift, if disrespected, is then withdrawn with dread effects for the world.

In previous writings, I contemplated the comingled meanings of green fire and Mother Earth.[3] Still, it is only now, writing for this series on the elements, that I realize a larger role of fire in Leopold's story. Along with the dying green fire, there is the firepower embodied in the rifle. These clashing fires derive from differing mythic origins—the first as gift of the Earth Mother, the second as stolen goods, purloined "from the gods" as in the well-known Greek story of Prometheus. That stolen fire demands outlet, inducing an intoxication, that "trigger-itch" Leopold acted on but later came to rue. Leopold's admonitory parable, then, speaks deeply to the apocalyptic folly of building a world powered by stolen fire.

Of all the elements, fire is the one that links up most readily with *power*, even forming that one word—*firepower*. The word is defined in a *Washington Times* photo essay for fans of firearms in this way: "The amazing firepower of the newest generation of guns: High-tech advances are boosting the firepower and lethality of guns—big and small."[4] Firepower is channeled not only into weaponry but also into machines, industry, and technology. As these suggest, the quest for firepower aligns with the characteristically Western aim to "conquer nature" and consequent catastrophes.

The desire for firepower leads to the wanton burning of fossil fuels, the prime agent of global warming, commonly understood as "a world on fire."[5] It spurs the development of ever-more-deadly personal firearms, military ordnance, and nuclear weapons (all of which serve as icons of masculinity). Firepower undergirds the ravaging consumerism—*consume* can mean "to burn with fire"—of the capitalist "good life." And the quest for firepower leads to the development of what its makers call "sparks of artificial general intelligence."[6] This new AGI, they suggest, might be superior to the human brain and poses an "extinction risk on par with nuclear war."[7] The word *extinction* derives from a root that means to "quench or put out." Stolen fire, in the misguided crusade to conquer nature, extinguishes green fire.

Even as the world burns, it remains easy to find exaltation of that crusade. For example, the atmospheric chemist and Nobel laureate Paul Crutzen and the journalist Christian Schwägerl hail the arrival of the new geological age of the Anthropocene. They claim it as the culmination of Western culture's hallmark "religious and philosophical idea—humans as the masters of planet Earth" and aver: "It's no longer us against 'Nature.' Instead, it's we who decide what nature is and what it will be."[8] A corrective to this absurdity is offered by writer and ritualist Malidoma Somé, who is of the Dagara ethnic group of Burkina Faso. He laments the ways that contemporary humanity, and the West in particular, manifests a spiritually broken relationship with fire, resulting in "the spilling of blood" as well as "the technological machine, which consumes nature around the world... consuming through both its speed and through the capitalist accumulation of land and rape of natural resources."[9] Such a relationship with fire is ordained by the nature of neither the element nor people. Fire, like all natural entities and forces, is both creative and destructive. The shapes that fire takes in our lives depend on cultural context, including foundational myths and rituals that guide human use.

The imperative, as Somé sees it, is to "make peace with fire," in ways ritualistic, spiritual, and practical. So, too, Leopold's vision suggests the need to bring the human relationship with fire back into balance with the green, with the processes of life (which include natural and accidental death but not species extirpation). There are many possible avenues toward such a goal. The one I take begins at the time of beginnings, considering mythic narratives about the origins of human relationship with fire and whether fire is understood as stolen power or as gift.

The Promethean Quest for Firepower (over Nature)

In a 1999 essay, the biologist and pyrographer par excellence Stephen Pyne stated that, historically, "the quest for fire is a quest for power." This is confirmed, he continues, in "virtually all fire-origin myths, where fire is "typically stolen by some culture hero."[10] Pyne is right to highlight the significance of myth, but this theme of stolen fire is not so universal as he suggests. Many other ancient stories suggest that fire is a gift from the Earth Mother. The righteous theft of fire is, however, the prevailing Western fire myth.

This story of Prometheus was first told around 700 BCE by Hesiod (whom Lewis Hyde names as the "earliest of misogynists").[11] In Hesiod's telling, Prometheus steals fire because he wants to bestow it on the "humans" he has made from clay so that they can acquire "civilization." I put the word *humans* in scare quotes because these progeny of Prometheus are all male. For Hesiod, as for Aristotle, the only real humans are men like themselves. I put *civilization* in scare quotes because, in colonialist and patriarchal reckoning, the attainment of civilization equates with society's success at dominating "nature" (with nature including those humans rendered socially inferior by being "defined into nature"—women, "savages," those who are enslaved, "freaks" and so on).[12]

The most significant rebuttal to the valorization of Prometheus and his theft of fire is Mary Shelley's great myth of the scientist who uses firepower in his quest to become a godlike creator of life. The novel's full title is *Frankenstein; or, The Modern Prometheus* (1818). Shelley mocks patriarchal paradigms by drawing upon the discourse of the sixteenth-century European scientific revolution to put words in the mouth of Dr. Frankenstein. Francis Bacon, the leading philosopher of that scientific revolution, favored gendered and sexually violent metaphors to describe the scientific method.[13] Echoing Bacon, Dr. Frankenstein describes himself as possessed by a "fervent longing to penetrate the secrets of nature." His method, he proclaims, is to pursue "nature to her hiding places" and force her to reveal those mysteries.[14] Centuries later, Shelley's critique of the rapist assault on nature and the theft of her secrets as well as the problematic Promethean myth remain largely disregarded in high-tech discourse. The authors of the 2014 tract *Regenesis: How Synthetic Biology Will Reinvent Nature and Ourselves* describe the field of synthetic biology as "the science of selectively altering the genes of organisms to make them do things that they wouldn't do in their original, natural, untouched state."[15] Meanwhile, celebratory references to stealing fire abound in such corporate names as Prometheus Biosciences and the Prometheus AI bot. The definitive biography of J. Robert Oppenheimer, the primary inventor of a nuclear weapon, is titled *The American Prometheus,* and that is intended as complimentary, invoking a tragic heroism.[16]

Returning to Hesiod, it turns out that the chief god Zeus (a liar and serial rapist of boys, girls, and women) orders a punishment for Prometheus's theft of fire. He must be bound to a rock to have his immortal liver devoured daily by an eagle. Zeus's punitive ire also extends to the clay men. He arranges for the manufacture of the novelty of Pandora, whom Hesiod describes as a "beautiful evil" and the first of the "deadly race and tribe of women."[17] Like Eve, Pandora is the bringer of ruin to men. She arrives bearing a jar containing multiple evils and plagues, which she opens and sets upon the world, with only hope remaining at the bottom.

Hesiod's influential story upends a preexisting oral tradition acknowledging that women have been around since the beginning and so has Pandora. Actually, Pandora is the beginning, for she is an Earth deity, a "mountain mother" whose name means "all giver." The classical scholar Jane Ellen Harrison understands Hesiod's version as a deliberate "patriarchal" inversion, whereby Zeus "takes over even the creation of the Earth-Mother who was from the beginning." In the earlier telling, Pandora was said to rise annually from the ground, holding a jar that she opens to bestow the gifts of the Earth, including, presumably, fire, to humanity.[18] In some versions, Pandora is a figure who can be either a man or a woman and who holds two jars, signifying the necessary message of all chthonic deities—all apparent opposites, including life and death and male and female, are really twin, necessary parts of the balanced dualism that is nature.[19] This is not to say there are only males and females. There are always people and states of being that inhabit liminalities. *Fire,* as marked in its complementary colors of red and green, is just such an emblem of the underlying cohesion of apparent opposites that is the terrestrial way.

In the past four centuries, Pandora's jar has become a "box" (slang for the vulva), reminiscent of other global myths. In one story from "Java's legendary past," a "flaming womb" manifests balance of female and male elements, a markedly nonpatriarchal understanding.[20] Other global myths hold that fire originally is "engendered magically in the genital organ of the sorceress."[21] In such stories, "older women 'naturally' possess fire in their genital organs and made use of it to do their cooking but kept it hidden from males, who were able to get possession of it only by trickery."[22] Such motifs suggest that the heroic theft of fire from Earth is a metaphor for the establishment of patriarchy and for rape, which is a form of breaking, entering, and stealing.

Seeing beyond Hesiod's disinformation, it is Pandora (and her jar or box), not Prometheus, who is the true bestower of fire. This recognition is underscored by the name of her daughter, Pyrrha, which translates to "of the fiery red earth."[23] In a Pandora-based "all giver" mythic paradigm, fire is a gift from Earth, one that requires a reciprocal exchange, a giving back. The plant biologist Robin Wall Kimmerer writes about the necessity of acknowledging that human beings are "showered every day with the gifts" of Mother Nature–Earth. "I've been told that my Potawatomi ancestors taught that the job of a human person is to learn, 'What can I give in return for the gifts of the Earth?'"[24] When I look at my own US culture, the return on Prometheus's theft of fire includes such horrors as gas-guzzling vehicles, automatic rifles, armaments including nuclear bombs and missiles, and greenhouse gases.

The Prometheus myth is thoroughly patriarchal, including in its association of fire with only men. Arguing in 1954 against the ideology of male supremacy, the anthropologist Evelyn Reed averred that it was women who "uncovered the uses of fire as a tool in their industries" and developed all the early industries, such as cooking, ceramics, and digging sticks.[25] In these activities, still coded as feminine, this quest for fire equates not with a quest for power as domination but with energy, creativity, community, and sustenance.

A decade after equating the quest for fire with the quest for power, Pyne does walk things back. He acknowledges, albeit without noting any factors of gender or ethnicity, that it is only "a small fraction" of humans who have been "responsible for releasing the combustion cascade that has washed over the planet." Pyne also reconsiders the dire impact of those myths about stealing power from the gods, as these define "anthropogenic fire as something wrenched from its natural setting, perhaps by violence, and redirected to human purposes."[26] This insight about the violence attending the birth of anthropogenic fire speaks mightily to why those who subscribe to the Promethean paradigm are now, in effect, reenacting that violence and burning down the Earth. The motif of stolen fire makes the element an object to be taken, rather than a living force, turning a source into a resource. This is the same sort of objectification that patriarchal cultures put on land, other than human beings, women, and people they enslave.

There is a crying need for alternative myths, symbols, rituals, and popular stories that point the way to a more balanced and sane relationship with fire. When the animated Disney film *Moana* appeared in 2016, some hailed it as conveying an ecofeminist message, with an eight-year-old Polynesian girl-hero facing off with apocalyptic fire and saving the world. Unfortunately, though, the story has fatal flaws. Primary among them is that *Moana* presents itself as Indigenous story, which, as the Ngati Porou scholar Tina Ngata declaims, it is not.[27] *Moana* actually tells a very Western story, one projecting its own rape of nature and disrespect of fire onto a colonized culture and then forgiving itself without making any actual change. Ironically, though, the Kānaka Maoli (Hawaiian) and Māori stories that *Moana* draws on and then distorts do give an alternative understanding of fire, specifically green fire as a gift of the earth.

The Gift of Green Firepower

Moana opens with sounds of chanting, then a word spoken in a Polynesian language, and then an older woman's voice introducing a cosmological tale with the words "In the beginning." The gathered children hear that woman, Moana's grandmother, tell a story about how the world was originally all water, until Te Fiti, Mother Island, arose from the deep. Te Fiti is young, beautiful, all green, and with a green heart bearing a spiral and radiating a light of its own. The grandmother affirms that this "heart has the greatest power ever known. It could create life," which Te Fiti generously shared with all the world. But, the grandmother laments, there were ungrateful ones who endeavored to steal Te Fiti's heart, for "they believed if they could possess it the great power of creation would be theirs."

This tale resonates deeply not with Indigenous stories, but with Western patriarchal origin myths that hinge on a heroic theft of fire, as well as rape and murder and the subsequent exploitation of the source or force of life. There is the upstart god Marduk in the *Enuma Elish*, the Babylonian creation story, who slays his grandmother, the creator Tiamat (in some versions killing her with his penis) and then forms the world from her dismembered body.[28] Ann Baring and Jules Cashford have averred that the story provides "the mythological roots of all three patriarchal [Abrahamic] religions," as evidenced in the biblical story of Yahweh defeating Leviathan.[29] There is also the regularly retold story of Perseus, who invades the cave (another womb symbol) of Medusa to chop off her snake-haired head and then use his prize as a weapon. And there is also, of course, Prometheus.

An overt reference to Prometheus shows up in the fake Indigenous story of Moana's grandmother when she introduces the actual Polynesian demigod Maui, whom Western interpreters regularly cast as an analogue of Prometheus, although he is not.[30] The grandmother tells of how Maui uses his sharp and pointy fishhook to break into the body of Te Fiti, prying out and stealing her heart.

Significantly, both the heart and the spiral are symbols of womb and vulva. Maui's implicit rape of Te Fiti causes her to crumble away. Feeling triumphant, Maui is making off with his prize when "confronted by another who sought the heart, Te Ka, a demon of earth and fire." Maui is no match for Te Ka, who knocks him from the sky, causing his fishhook and the green-fire heart to be lost to the ocean. The grandmother continues: "Even now, a thousand years later, Te Ka and the demons of the deep still continue to hunt for the heart, hiding in the darkness that will continue to spread, chasing away our fish, draining the life from island until every one of us is devoured by the jaws of ravenous death." There is hope, though. The grandmother ends with a prophecy that there will be a savior who will find Maui and work with him to restore the heart to Te Fiti and the radiant greening power to the world.

That savior is Moana (a name that means "ocean"), the daughter of a chief. Her people are in deep trouble as the blight the grandmother describes has come to their island. But the ocean has chosen Moana as the savior and delivers the heart to her on a wave. Moana sets off on the ocean against her father's command, finds Maui, induces him to work with her, and manages to retrieve his fishhook. After many trials, they arrive at the Mother Island, only to encounter Te Ka—red, fiery, molten, raging, volcanically erupting. All seems lost, but Moana sees a spiral configuration in the heart zone of Te Ka. She holds up the shimmering green heart and returns it to Te Ka. At this, Te Ka disappears and beauteous green Te Fiti reappears. She smiles, places her hand on the rock, and transforms it into flowering life. Maui, prodded by Moana, apologizes to Te Fiti, who forgives him without much fuss and even repairs the instrument of her rape, his fishhook. Moana triumphantly returns to her now healthy and fruitful island and will be the next chief.

With a theme of threatening eco-apocalypse, *Moana* alludes to the devastation happening in the South Pacific, notably climate change. Yet the film attributes this not to the effects of settler colonialism, nuclear bomb testing, militarization, consumerism, and tourism. Instead, it attributes the collapse to an original sin by a sacred hero in Polynesian culture while also making it seem as if the story being told were an actual Polynesian myth. Not only is this a kind of theft or cultural appropriation; Indigenous stories are not meant for entertainment. They are conduits of philosophy, cosmology, ethics, and instructions for survival and continuance. For example, the Kānaka Maoli scholar Noenoe Silva explains that *moʻolelo* are "a major way we make communal, transcendent meaning out of human experience." She stresses the harms of colonialist distortion of "ancestral stories" and calls for their liberation from that "captured state," for it is "through the moʻolelo that we may clearly understand ourselves as linked to our ancestors and our land."[31] Stealing Indigenous stories, then, is part of a continuing colonization, causing cultural erasure and the disconnection that allows for land to be further taken and abused.

Disney, perhaps aware that it might be accused of cultural appropriation, had arranged for the creation of an entity called the Ocean Story Trust to advise its writers. Perhaps because of these advisers' contributions, the Filipino-Pohnpeian scholar Vicente Diaz notes, there are key allusions in the film to "substantive cultural material that approaches the spiritual and the sacred."[32] But as he and many others point out, the way that *Moana* incorporates these elements terribly distorts their philosophy and spirituality.

This is apparent in the figure of Te Ka, the "lava demon," who direly misrepresents the reality of the beloved Hawaiian volcano goddess Pele. As the Native Hawaiian scholar ku'ualoha ho'omanawanui explains in her book *Voices of Fire: Reweaving the Literary Lei of Pele and Hi'iaka*, the name Pele means "lava eruption," and the goddess can appear as red, molten lava that is both Earth devouring and Earth making. This is how the Hawaiian islands were, as she puts it, "born in fire" millennia ago.[33] *Moana*'s Te Ka is suggestive of Pele, but Te Ka is represented as a dangerous demon who must be put down, her molten redness cast as negatively old and ugly, the inverse of benevolent green, beauteous, and youthful Te Fiti. This splitting of the Earth Mother is typical of the patriarchal West; Mother Nature–Earth is "good" when virginal, passive, young, lovely, forgiving, and life giving, and "bad" when old, active, and death giving. This framework of oppositional hierarchy—including male over female, culture over nature, civilization over savagery—works to provide the basis, as the ecofeminist philosopher Val Plumwood argues, for male and white supremacist ideologies and justification for patriarchal men's crusade to master nature.[34]

Kānaka Maoli ontology as expressed through narratives about Pele is utterly different. In these, Pele is tied to another female figure: "Pele has a sister, Hi'iaka with whom she regularly appears." There is an ecological meaning to this configuration, as Pele's sister provides "a balancing force to Pele's temperamental nature." As such, ku'ualoha ho'omanawanui explains, "The two are 'ēko'a (oppositional and complementary) forces; together their pono (counterbalanced) relationship signifies a foundational tenet of Hawaiian ontology."[35] *Moana* alludes to this pairing, but then undoes its ecological significance. Te Fiti and Te Ka *never* appear together. Indeed, the happy ending is assured only when "bad" Te Ka disappears and "good" Te Fiti comes back to work a little magic and absolve her violator before passively settling back into the land.

The representation of Maui also disregards, if not fully upends, ecological meanings, including those regarding the origin of fire. In Polynesian sacred narrative, Maui is a culture hero and demigod. He surfaces islands from the deep with his fishhook, brings fire, and makes humans human. In *Moana*, Maui is cast as a wisecracking, American-type action hero—hypermasculine, egocentric, and immature. And in the stories that Indigenous people tell, he does not steal fire but brings it after learning to honor it as a gift from the Mother.

One Māori story tells how the great goddess "Hina, in the form of an 'alae'ula, mudhen bird,' teaches him the great secret of making fire."[36] Another Māori story features the fire goddess Mahuika, with whom all fire originates. She, like Pandora, is a mountain mother. The story begins when Maui, while staring into the fire, gets curious about where fire comes from. To find out, Maui deliberately extinguishes the fires in a village so that he will be asked to journey to the dwelling of Mahuika and ask her to make a gift of fire to humans. Maui gets instructions and directions to her abode from his mother (Mahuika is his grandmother) and sets off. After his arrival, Mahuika listened carefully to Maui, and then she laughed. She pulled a fingernail from one of her burning fingers and gave it to him. "Take this fire as a gift to your people." She also gave him an instruction: "Honor this fire as you honor me."

Maui leaves with the gift of fire, a fire that is the essence of the goddess, part of her body-being. But he fools around, throws the fingernail to the ground to watch the fire explode, and then goes back to request still another gift of fire, a pattern he repeats. Mahuika indulges him for a while with more fingernails and then toenails, but finally gets fed up and throws down a penultimate toenail, engulfing Maui in fire. The demigod shape-shifts into a flying hawk to try to escape and pleads to Tāwhirimātea, the god of weather, to intervene with rain, which he does. Mahuika, though, throws down her last toenail, and with that, her presence in the world is "diminished." Maui (and humans) will no longer have access to the primal fire. But with luck, some of that toenail fire "flew into the trees, planting itself in the Mahoe tree, the Tōtara, the Patete, the Pukatea, and the Kaikōmako trees. These trees cherished and held onto the fire of Mahuika, considering it a great gift."[37] Because of the green trees' honoring of the gift, Maui can return to the people, not with primal fire, but with what we might recognize as a form of green fire—the dry wood from those green trees. He can teach everyone how to release the fire by rubbing the sticks together. As an outsider, I cannot know the full meanings of this story. One clear lesson, though, is that fire is a living force and a gift from the source, not an object, thus necessitating respect for both the giver and the gift. This is something the trees knew and practiced, to the benefit of humanity. Their gift to humans is quintessentially green fire.

The immaturity, egotism, and pumped-up hypermasculinity that *Moana* projects onto Maui are Western culture's own traits. When will we mature enough to know that the chthonic forces of Earth are complementary, that life and death are "twin beings, gifts of our Mother, the Earth," and that fire and greenness are kin, such that humans must act to keep them in balance by always seeking to green firepower?[38]

However flawed the film, some viewers see beyond and make their own meanings. I take heart from an expression of fan art sold on Etsy.[39] A round, silver pendant shows Te Fiti and Te Ka together, inseparable, joined in the embrace of a double spiral. One swirling strand is green and moves out from the center to end in a dreaming Te Fiti. The other, a black strand, ends in the red, fiery, and wide-awake Te Ka. Te Ka's hands rest on the green spiral, while Te Fiti's hands rest on the black. Here is a vision of a green firepower, a mythic alternative so necessary for our world on fire.

notes

1. Aldo Leopold, "Thinking Like a Mountain," in *Sand County Almanac: With Essays on Conservation from Round River* (New York: Ballantine Books, 1949), 137–41.
2. Aldo Leopold, "The Virgin Southwest," in *The River of the Mother of God and Other Essays by Aldo Leopold*, ed. Susan L. Falder and J. Baird Caldicott (Madison: University of Wisconsin Press, 1991), 181.
3. Jane Caputi, *Call Your "Mutha'": A Deliberately Dirty-Minded Manifesto for the Earth Mother in the Anthropocene* (New York: Oxford University Press, 2020).
4. "The Amazing Firepower of the Newest Generation of Guns" (photo gallery), *Washington Times*, https://www.washingtontimes.com/multimedia/collection/guns-future-amazing-firepower/.
5. Editorial Board, "Scenes from a World on Fire," *New York Times*, December 31, 2021, https://www.nytimes.com/2021/12/31/opinion/climate-change-glasgow-united-states.html.
6. Sébastien Bubeck, Varun Chandrasekaran, Ronen Eldan, Johannes Gehrke, Eric Horvitz, Ece Kamar, Peter Lee, Yin Tat Lee, Yuanzhi Li, Scott Lundberg, Harsha Nori, Hamid Palangi, Macro Tulio Ribeiro, and Yi Zhang, "Sparks of Artificial General Intelligence: Early Experiments with GPT-4" (Microsoft Research, 2023), https://arxiv.org/pdf/2303.12712.pdf.
7. Sheila Chiang, "Advanced AI," *CNBC*, May 20, 2023, https://www.cnbc.com/2023/05/31/ai-poses-human-extinction-risk-sam-altman-and-other-tech-leaders-warn.html.
8. Paul J. Crutzen and Christian Schwägerl, "Living in the Anthropocene: Toward a New Global Ethos," *Yale Environment 360*, January 24, 2011.
9. Malidoma Somé, "A Year to Make Peace with Fire," *Moon Magazine*, 2017, http://moonmagazine.org/malidoma-some-a-year-to-make-peace-with-fire-2017-03-04/.
10. Stephen J. Pyne, "Consumed by Either Fire or Fire: A Prolegomenon to Anthropogenic Fire," in *Earth, Air, Fire, Water*, ed. Jill Kerr Conway, Kenneth Kenison, and Leo Marx (Amherst: University of Massachusetts Press, 1999), 78–101, esp. 80.
11. Lewis Hyde, *Trickster Makes This World: Mischief, Myth and Art* (New York: Farrar, Straus & Giroux, 1998), 38.
12. Maria Mies, *Patriarchy and Accumulation on a World Scale*, 2nd ed. (New York: Zed Books, 1999), 75.
13. Carolyn Merchant, *The Death of Nature: Women, Ecology and the Scientific Revolution* (San Francisco: HarperSanFrancisco, 1980).
14. Mary Shelley, *Frankenstein; or, The Modern Prometheus* (1818; New York: Dover, 1994), 21, 33.
15. George Church and Ed Regis, *Regenesis: How Synthetic Biology Will Reinvent Nature and Ourselves* (New York: Basic Books, 2014), 143.

16. Kai Bird and Martin J. Sherwin, *American Prometheus: The Triumph and Tragedy of J. Robert Oppenheimer* (New York: Alfred A. Knopf, 2005).
17. Hesiod, *Theogony*, https://www.theoi.com/Text/HesiodTheogony.html.
18. Jane Ellen Harrison, *Mythology* (1923; New York: Harcourt, Brace and World, 1963).
19. Ann Baring and Jules Cashford, *The Myth of the Goddess: Evolution of an Image* (New York: Arkana Penguin Books, 1991), 5528.
20. Barbara Watson Andaya, *The Flaming Womb: Repositioning Women in Early Modern Southeast Asia* (Honolulu: University of Hawai'i Press, 2006), 1.
21. Mircea Eliade, *The Forge and the Crucible*, trans. Stephen Corrin (Chicago: University of Chicago Press, 1962), 40.
22. Eliade, 6.
23. Martha Ann and Dorothy Myers Imel, *Goddesses in World Mythology* (Santa Barbara, CA: ABC-CLIO, 1993), 211.
24. Robin Wall Kimmerer, "Returning the Gift," *Minding Nature* 7, no. 2 (Spring 2014), https://www.humansandnature.org/returning-the-gift.
25. Evelyn Reed, "The Myth of Women's Inferiority," *Fourth International* 15, no. 2 (Spring 1954): 58–66.
26. Stephen J. Pyne, *The Pyrocene: How We Created an Age of Fire and What Happens Next* (Oakland: University of California Press, 2021), 146.
27. "Despite Claims of Authenticity, Disney's *Moana* Still Offensive" (interview with Tina Ngata), *Rising Up with Sonali*, November 23, 2016, https://risingupwithsonali.com/despite-claims-of-authenticity-disneys-moana-still-offensive/.
28. Michelle I. Marcus, "Sex and the Politics of Female Adornment in Pre-Achaemenid Iran (1000–800 BCE)," in *Sexuality in Ancient Art: Near East, Egypt, Greece, and Italy*, ed. Natalie Boymel Kampen (Cambridge: Cambridge University Press, 1996), 50.
29. Baring and Cashford, *Myth of the Goddess*, 275.
30. David Leeming, *The Oxford Companion to World Mythology* (New York: Oxford University Press, 2005), 137.
31. ku'ualoha ho'omanawanui, *Voices of Fire: Reweaving the Literary Lei of Pele and Hi'iaka* (Minneapolis: University of Minnesota Press, 2014), xxvii.
32. Vicente Diaz, "Don't Swallow (or Be Swallowed) by Disney's Culturally Authenticated Moana," *Indian Country News*, November 13, 2016.
33. ho'omanawanui, *Voices of Fire*.
34. Val Plumwood, *Feminism and the Mastery of Nature* (New York: Routledge, 1993).
35. ho'omanawanui, *Voices of Fire*, xxvii–xxviii.
36. Tēvita O. Ka'ili, "Goddess Hina: The Missing Heroine from Disney's Moana," *Huffington Post*, November 26, 2016, https://www.huffpost.com/entry/goddess-hina-the-missing-heroine-from-disney%CA%BCs-moana_b_5839f343e4b0a79f7433b6e5.
37. Wiremu Grace, "How Maui Brought Fire to the World," *Te Kete Ipurangi* (New Zealand Ministry of Education), https://eng.mataurangamaori.tki.org.nz/Support-materials/Te-Reo-Maori/Maori-Myths-Legends-and-Contemporary-Stories/How-Maui-brought-fire-to-the-world.
38. Paula Gunn Allen, "The Woman I Love Is a Planet, the Planet I Love Is a Tree," in *Reweaving the World: The Emergence of Ecofeminism*, ed. Irene Diamond and Gloria Feman Orenstein (San Francisco: Sierra Club Books, 1990), 52–57, esp. 52.
39. Disney Bounded, "Disney Te Fiti Te Ka Heart of the Ocean Princess Moana Demigod Polynesian Hawaiian Fish Extender Silver Pendant Necklace Jewelry," https://www.etsy.com/listing/552959197/disney-te-fiti-tefiti-te-ka-teka-heart.

Permissions

These credits are listed in the order in which the relevant contributions appear in the book.

"Vestment" by Jane Hirshfield, from *The Asking: New & Selected Poems* (New York: Knopf, 2023). Copyright © 2023 by Jane Hirshfield. Used by permission of the author, all rights reserved.

"Instructions for a Wampanoag Clambake" by Lucille Lang Day, from *Becoming an Ancestor: Poems* (West Somerville, MA: Červená Barva Press, 2015). First published in *The Tower Journal*. Reprinted with permission of the author.

"Simile" by David Baker, from *Scavenger Loop: Poems*. Copyright © 2015 by David Baker. Used by permission of W. W. Norton & Company, Inc.

"I Vow to Be the Small Flame" by Tamiko Beyer, from *Last Days* (New Gloucester, ME: Alice James Books, 2021). Used by permission of Alice James Books.

"Let Me" by Camille T. Dungy first appeared in *The New Yorker* (April 12, 2021).

Acknowledgments

Our gratitude runs deep for the community of kin who made this series possible. Strachan Donnelley, the founder of the Center for Humans and Nature, was animated and inspired by big questions. He liked to ask them, he enjoyed following the intellectual and actual trails where they might lead, and he knew that was best done in the company of others. Because of this, and because Strachan never tired of discussing the ancient Greek philosopher Heraclitus, who was partial to Fire, we think he would be pleased by the collective journey represented in *Elementals*. One of Strachan's favorite terms was "nature alive," an expression he borrowed from the philosopher Alfred North Whitehead. The words suggest activity, vivaciousness, generous abundance—a world alive with elemental energy: Earth, Air, Water, Fire. We are a part of that energy, are here on this planet because of it, and the offering of words given by our creative, empathic, and insightful contributors is one way that we collectively seek to honor *nature alive*.

A well-crafted, artfully designed book can contribute to the vitality of life. For the mind-bending beauty of the cover design, cheers to Mere Montgomery of LimeRed; she is a delight to work with and LimeRed an incredible partner in bringing to visual life the Center for Humans and Nature's values. For an eye of which an eagle would be envious, a thousand blessings to the deft manuscript editor Katherine Faydash. For the overall style and subtle touches to be experienced in the page layout and design, we profoundly thank Riley Brady. We also wish to thank Ronald Mocerino at the Graphic Arts Studio Inc. for his good-natured spirit and attention to our printing needs, and Chelsea Green Publishing for being excellent collaborators in distribution and promotion.

Thank you to our colleagues at the Center for Humans and Nature, who are elemental forces in their own rights, including our president Brooke Parry Hecht, as well as Lorna Bates, Anja Claus, Katherine Kassouf Cummings, Curt Meine, Abena Motaboli, Kim Lero, Sandi Quinn, and Erin Williams. Finally, this work could not move forward without the visionary care and support of the Center for Humans and Nature board, a group that carries on Strachan's legacy in seeking to understand more deeply our relationships with *nature alive*: Gerald Adelmann, Julia Antonatos, Jake Berlin, Ceara Donnelley, Tagen Donnelley, Kim Elliman, Charles Lane, Thomas Lovejoy, Ed Miller, George Ranney, Bryan Rowley, Lois Vitt Sale, Brooke Williams, and Orrin Williams.

—Gavin Van Horn and Bruce Jennings
series coeditors

Stephanie Krzywonos would like to thank Gavin for inviting her to be one of the "elements" in this collective project, Bruce and Gavin for their editorial guidance, Nickole and Craig for gathering such stunning poems for the *Fire* volume, and the wise group of writers who generously shared their fiery insights. She would also like to thank every animating beam of light—sunlight, campfire light, twilight—that shows us the way.

Contributors • volume iv

Glenn A. Albrecht is Honorary Associate in the School of Geosciences, University of Sydney, New South Wales, Australia. He retired as Professor of Sustainability, Murdoch University, in 2014. He continues to work as an environmental philosopher and published a book, *Earth Emotions* (Cornell University Press, 2019). *Earth Emotions* was published in French and Spanish in 2020. Dr. Albrecht has developed the theme of the psychoterratic (psyche-earth), or negative and positive emotional states connected to the state of the Earth. His new concepts, such as the Symbiocene, are now well established in the international scholarly literature and creative arts.

David Baker has published many books of poetry and prose about poetry, most recently *Whale Fall* (W. W. Norton, 2022), *Swift: New and Selected Poems* (Norton, 2019), and *Seek After: On Seven Modern Lyric Poets* (Stephen F. Austin, 2018). He writes frequently about ecology and the lyric and continues to edit "Nature's Nature," an annual ecopoetry folio, for the *Kenyon Review*. He lives in Granville, Ohio.

Photo by Susi Franco

Tamiko Beyer (she/her) is the author of *Last Days* (Alice James Books, 2021) and *We Come Elemental* (Alice James Books, 2013), and co-editor of *Poetry as Spellcasting* (North Atlantic Books, 2023). She has received awards from Lambda Literary, PEN America, and the Astraea Lesbian Writers Fund, and fellowships and residencies from Kundiman, Hedgebrook, and VONA, among others. She publishes *Starlight & Strategy*, a monthly newsletter for shaping change. A social justice communications writer and strategist, she spends her days writing truth to power. She lives and writes on Massachusett land with her human, canine, and plant family.

Eiren Caffall (she/her) is a writer and musician based in Chicago, Illinois. Her writing on loss and illness, oceans, and extinction has appeared in *Guernica, Los Angeles Review of Books, Al Jazeera, Literary Hub, Minding Nature,* and *The Rumpus.* She has been the recipient of a *Social Justice News Nexus* fellowship in environmental journalism at Northwestern University's Medill School of Journalism and a *Frontline: Environmental Reportage* residency at the Banff Centre for the Arts. She is the screenwriter for the short film *Becoming Ocean.* She lives in the Logan Square neighborhood with her husband and son.

Jane Caputi (she/her) is Professor of Women, Gender, and Sexuality Studies at Florida Atlantic University. Her most recent book is *Call Your "Mutha'": A Deliberately Dirty-Minded Manifesto for the Earth Mother in the Anthropocene* (Oxford University Press, 2020). She also made the short documentary *Feed the Green: Feminist Voices for the Earth* (2016), distributed by Women Make Movies.

Photo by
Maude Roxby

Rina Garcia Chua (she/her/siya) is a creative and critical scholar from the Philippines who is currently based in British Columbia, Canada. Her poems have been previously published in numerous journals, some of which are *World Literature Today, Asteri(x), g u e s t,* and *The Goose: Journal of Arts, Culture, and Environment in Canada,* of which she is currently coeditor. Rina is completing her poetry chapbook, *A Geography of (Un)Natural Hazards,* which is a visual and poetic response to migrant and arriving cultures, liminal environments, and violences of form and language. She can be reached at her website: http://rinagarciachua.com

Catharina Coenen (she/her) is a first-generation German immigrant to Northwest Pennsylvania, where she teaches biology at Allegheny College. Her essays have appeared in Best-of-the-Net, *Threepenny Review, American Scholar, Christian Science Monitor,* and elsewhere. Catharina is the recipient of a Creative Nonfiction Prize from *The Forge,* the *Appalachian Review*'s Denny Plattner Creative Nonfiction Prize, a Creative Nonfiction Foundation Science-as-Story Fellowship, a Fulbright Fellowship, and a Hedgebrook Residency.

Lucille Lang Day (she/her) is the author of four poetry chapbooks and seven full-length collections, including *Birds of San Pancho and Other Poems of Place.* She also edited *Poetry and Science: Writing Our Way to Discovery,* coedited *Fire and Rain: Ecopoetry of California* and *Red Indian Road West: Native American Poetry from California,* and authored two children's books and a memoir. Her many honors include the Blue Light Poetry Prize, two PEN Oakland–Josephine

Miles Literary Awards, the Joseph Henry Jackson Award, and eleven Pushcart nominations. The publisher of Scarlet Tanager Books, she is of Wampanoag, British, and Swiss-German descent. https://lucillelangday.com

Charlotte Du Cann is a writer, editor, and codirector of the Dark Mountain Project. Her most recent book, *After Ithaca: Journeys in Deep Time*, is about recovering our core relationships with the mythos and sentient Earth, revolving around the underworld tasks of Psyche. She teaches ensemble creative practice and lives on the wild salty edge of East Anglia. charlotteducan.net

Camille T. Dungy (she/her) is the author of the book-length narrative *Soil: The Story of a Black Mother's Garden*. She has also written four collections of poetry, including *Trophic Cascade* and the essay collection *Guidebook to Relative Strangers*. She edited *Black Nature: Four Centuries of African American Nature Poetry* and coedited the *From the Fishouse* poetry anthology. Dungy is Poetry Editor for *Orion* magazine. A University Distinguished Professor at Colorado State University, Dungy's honors include the 2021 Academy of American Poets Fellowship, a 2019 Guggenheim Fellowship, an American Book Award, and fellowships from the National Endowment for the Arts in both prose and poetry.

José G. González (he/him) is a Chicano educator with experience as a K-12 public education teacher, environmental education advisor, outdoor science education instructor, and university adjunct faculty. He resides in Sacramento, California.

The Honorable Ron W. Goode has been Tribal Chairman of the North Fork Mono Tribe since 1983. He is a veteran of the US Army, and Life Member of the Sierra Mono Museum and the US Judo Federation. Ron holds a sixth-degree black belt in judo and still enjoys teaching. He is also a retired community college professor in ethnic studies. Ron was inducted in the Clovis Hall of Fame for his work in education and community service in 2002. In 2006 he was selected California Indian Education Teacher of the Year. In 2013, Ron was recognized by Governor Brown for his work with the Department of Water Resources and as cofounder of the Tribal Water Summit. In 2022, Ron was honored by the Society of California Archaeology for the Lifetime Achievement Award for his work in cultural preservation.

Photo by Curt Richter

Writing "some of the most important poetry in the world today" (*New York Times Magazine*), **Jane Hirshfield** is the author most recently of *The Asking: New & Selected Poems* (Knopf, 2023); two collections of essays; and four books collecting and co-translating world poets from the deep past. Hirshfield's honors include the Poetry Center Book Award, the California Book Award, and finalist selection for the National Book Critics Circle Award. Translated into seventeen languages, Hirshfield is a former chancellor of the Academy of American Poets and an elected member of the American Academy of Arts & Sciences.

Stephanie Krzywonos (she/her) is a Xicana writer and graduate student at the University of Iowa's Nonfiction Writing Program. Her first book, *Ice Folx: An Antarctic Memoir*, is forthcoming from Atria in 2025.

Miriam Morrill is a fire education specialist and illustrator who has provided training and support to various fire-planning efforts and collaboratives in the United States, Australia, the Republic of Palau, and Jamaica, as well as the International Association of Wildland Fire. Her illustration work has been used in a number of fire and smoke research projects and fire plans in the United States and Canada. After retiring from a twenty-seven-year career in wildland fire mitigation and wildlife management, Miriam developed a program called Pyrosketchology to help people form a deeper awareness and understanding of fire through a nature journaling practice.

Nina Mingya Powles (she/her) is a poet, zine-maker, and librarian from Aotearoa New Zealand, currently living in London. She is the author of several poetry pamphlets, zines, and poetry books, most recently *Magnolia* 木蘭 and a collection of essays, *Small Bodies of Water*.

Em Strang (she/her) is a poet, novelist, creative mentor, and founder of the Scottish charity Three Streams. Her writing preoccupations are with nature, spirituality, and the question of evil. Em's first full collection of poetry, *Bird-Woman*, was published by Shearsman in October 2016, was shortlisted for the Seamus Heaney Best First Collection Prize, and won the 2017 Saltire Poetry Book of the Year Award. Her second collection, *Horse-Man*, was published in September 2019 and was shortlisted for the Ledbury Munthe prize for Best Second Collection. Her first novel, *Quinn*, was shortlisted for the 2019 Fitzcarraldo Editions Novel Prize and was published by Oneworld in 2023.

Isaac Yuen's (he/him) creative work has been published in *AGNI, Gulf Coast, Orion, Pleiades, Shenandoah, Tin House,* and other literary publications. Winner of a Pushcart Prize, his debut nature essay collection titled *Utter, Earth* from West Virginia University Press weaves together the human and more-than-human world with wordplay and earthplay. A first-generation Hong Kong–Canadian, Isaac is currently a writer-in-residence at the HWK Institute of Advanced Studies in Delmenhorst, Germany.

Tyson Yunkaporta is an Aboriginal scholar, founder of the Indigenous Knowledge Systems Lab at Deakin University in Melbourne, and author of *Sand Talk.* His work focuses on applying Indigenous methods of inquiry to resolve complex issues and explore global crises.